Dec 1999

To Sandy —

From One Wine Lover to
Another!

Enjoy the Book!
Enjoy the Wine!

Steven Ko

A Sense *of* Place

A Sense *of* Place

AN INTIMATE PORTRAIT
OF THE NIEBAUM-COPPOLA WINERY
AND THE NAPA VALLEY

S TEVEN K OLPAN

FOREWORD BY FRANCIS FORD COPPOLA

Routledge
NEW YORK LONDON

Published in 1999 by

Routledge
29 West 35th Street
New York, NY 10001

Published in Great Britain by

Routledge
11 New Fetter Lane
London EC4P 4EE

Color photos courtesy of Steven Kolpan
Black and white photos courtesy of the Niebaum-Coppola Winery

Interior text design by Jeff Miller Stratford Publishing Services

Printed in the United States of America on acid free paper

10 9 8 7 6 5 4 3 2 1

Library of Congress Cataloging-in-Publication Data
Kolpan, Steven.
A Sense of place : an intimate portrait of the Niebaum-Coppola Winery and the Napa Valley / by Steven Kolpan.
p. cm.
Includes bibliographical references and index.
ISBN 0-415-92004-3
1. Wine and winemaking—California—Napa Valley. 2. Wineries—California—Napa Valley. I. Title.
TP557.K65 1999
641.2'2794'19—dc21 99-35012
CIP

For my parents,
Ruth and Jacob

Contents

Acknowledgments

*T*have many people to thank for making this book a reality, and I am happy to have the opportunity to do so. Each one, in his or her own way, has contributed to making my creative life richer. I could not have completed *A Sense of Place* without their help, encouragement, and love.

To Suzanne Hamlin, the finest food and wine journalist in the English language and my life partner, thank you for introducing me to Francis and Eleanor Coppola, and for encouraging me to undertake this project.

To Melissa Rosati, Publishing Director at Routledge, who loved the idea for this book, and by her nurturing of that idea, refined my work and kept me on track and on schedule. Every author should be so fortunate.

Special thanks to Kathleen Talbert and Talbert Communications for presenting my idea for this book to Francis and Eleanor Coppola, and for providing valuable assistance at every critical point in the development of *A Sense of Place.*

Thanks to The Culinary Institute of America in the Hudson Valley and at Greystone in the Napa Valley. Granting me a funded sabbatical leave to research and write was extraordinarily gracious and generous. I want to especially thank Ferdinand Metz, President of The Culinary Institute of America, and Tim Ryan, Executive Vice President. Also, many thanks to Fred Mayo, Bill Reynolds, and Ezra Eichelberger for supporting my sabbatical project, and to professors Michael Weiss, Brian Smith, and Ralph Johnson for teaching wine classes at the Institute, including mine, during my sabbatical. To Henry Woods, Mary Cowell, Lorna Smith, and Liz Johnson, thanks for providing film, lenses, and technical assistance. To Eve Felder, Laura Pensiero, Gianni Scappin, and Gary Allen, thanks for listening.

Writing *A Sense of Place* required me to travel about 17,000 miles, all of it in my old Volvo. Thanks to Charlie Rascoll, Debbie Howe, and Lucy Howe-Rascoll for a house to call home; to Ron and Audrey Martinez, and to Isabella Siena Martinez, my goddaughter, for waiting until I got to California for Bella to be born, and to Reuben Katz and Janice Cimberg for their never-ending Napa Valley hospitality.

For keeping the home fires burning and the lights on in Woodstock, thanks to my best friend, Bob Schuler, and to Risa Levine. For legal help above and beyond the call of what's reasonable, thanks to Paul Kellar.

Thank you to the staff of the extraordinary Napa Wine Collection at the St. Helena Library for their tireless and patient assistance. This book would not have been possible without unlimited access to the library's holdings.

Thanks to Jeremy Benson for inviting me to the eye-opening Rutherford Dust Society tasting, and to all the Rutherford wine producers for their generosity with time and wine.

To Robin Lail and to Dennis Fife, thank you for your cooperation and for your trust.

To everyone at the Niebaum-Coppola estate, many thanks. To Anaheid for actively helping my research; to Bruce Dukes for his amazing grasp of the science of wine and his Aussie humor; to Erle Martin, who explained how the place ticks, and always had time to answer my questions. Special thanks to Scott McLeod, winemaker, for spending innumerable days with me in the vineyards and for making sure that I got everything I needed to write this book; and to Rafael Rodriguez, for sharing and *being* history, for his insights, his wisdom, his humor, and above all, his integrity.

Finally, thank you to Francis and Eleanor Coppola, who allowed me totally free access to the Niebaum-Coppola estate and archives, to their home, their ideas, and their dreams. True to their word, the Coppolas never asked for any editorial oversight of *A Sense of Place.* I thank them for that and for the trust they have shown in me by encouraging me to present the story of Inglenook/Niebaum-Coppola in my own words.

Steven Kolpan
Woodstock, New York
June 2, 1999

Foreword
A Sense of Heritage

*M*y adventure with the Napa Valley began in the early '70s. We were coming back from making *The Rain People,* and my young assistant George Lucas, who was from Modesto, suggested we drive back through northern California, through the Napa Valley. I had always heard of the area, because my grandfather made wine from Napa Valley grapes, and I pictured this as a little Italian community in the middle of the fields of grapes.

A few years later, after our family had moved to San Francisco, my wife and I thought it might be nice to have a little house and maybe an acre or two of grapes so that we could make wine in the Italian-American tradition. We came and looked for cottages and the realtor told us they were going to auction off the Niebaum estate, part of the great Inglenook. He said, "It isn't for you, but it would be fun to see it." At the end of the road was this stunning Victorian mansion on this enormous wine estate. It was like that George Stevens film, *A Place in the Sun,* with rich people sitting around a pool and a Mercedes out front. To anyone not raised with these things it was unbelievable. This was what everyone considered the queen of the Napa Valley, perhaps *the* great American château. We ended up being able to buy it, but it was a far cry from that little cottage we thought we wanted.

As I began to learn about the heritage of the estate, I realized that we had — by accident — come into possession of something extraordinary. It was like a family who inherits a racehorse and realizes it would be absurd not to race such a thoroughbred. We quickly realized that it was crazy not to make wine from this legendary property.

We had the privilege of tasting some very old Inglenook wines and could taste the greatness of the estate. Buying the rest of the property in 1995 and restoring the estate to its historic dimension is truly a dream come true. I didn't care that the Inglenook name didn't come with the purchase. It had been irreparably damaged anyway. What I wanted most was the heritage that makes the property unique — Gustave Niebaum's legacy, which was the creation of a world-class wine estate — the vineyards, of course, that grand château, the collection of old Inglenook wines, and all the other medals, awards and memorabilia that are now in our museum. Heritage is everything. If you have the heritage and respect it, it is an endless source of inspiration.

In America, very few things that are split apart are ever put together again. There seems to be no incentive to respect heritage and tradition. My family and I have vowed that this place will never be split up or sold again. In *A Sense of Place: An Intimate Portrait of the Niebaum-Coppola Winery and the Napa Valley,* Steven Kolpan tells the stories behind this estate's founding, ascent to greatness, eventual dissolution, and final reunification with great care, insight, and passion for this irreplaceable heritage.

When you have the land, you have the grapes, and when you have the grapes, you have the wine. This land has produced world-class wines for over 100 years, so if we can continue to do that under our stewardship, it would be a tremendous achievement. I've lived here for 25 years, surrounded by my family and beautiful nature. There are mushrooms on the mountains, vegetables in the garden, and fruit on the trees within arm's reach. This place is truly heaven on earth.

Eleanor and I invite you, the reader, to visit the Niebaum-Coppola Estate Winery—one of America's great wine estates, whole again.

Francis Ford Coppola
Rutherford, California
May 10, 1999

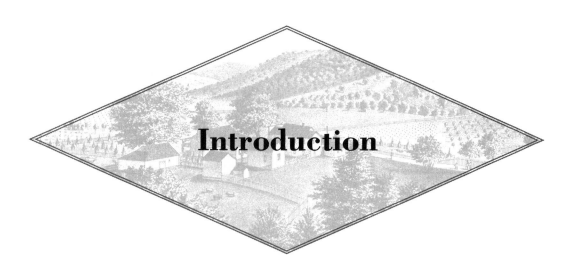

Introduction

*I*n the fall of 1995, while I was teaching a wine course at Greystone, the Napa Valley campus of The Culinary Institute of America, I was invited to have dinner at the Niebaum-Coppola Estate with a dozen food and wine journalists and the owners of the winery, Francis and Eleanor Coppola.

Niebaum-Coppola is about six miles south of Greystone, straight down Highway 29. Unlike Greystone, a former Christian Brothers winery and imposing almost-roadside stone castle, the old Niebaum estate, anchored by the original Inglenook stone château (and designed by the same architect as Greystone, William Mooser), is out of sight, well off the highway down a long straight road.

After parking the car, we walked up the hill to the château and were greeted by Eleanor Coppola, and somewhat diffidently, by Francis Coppola, who, wearing his trademark Hawaiian shirt and propelled by air-cushioned sneakers, seemed anxious to show us around. We took the whirlwind tour.

We saw the plans for the public space in the old winery building, admired a perfectly restored Tucker automobile—a souvenir of the film Francis Coppola made about a dreamer—and the old Coppola family photographs juxtaposed with Niebaum memorabilia. Even though it was late in the day, carpenters were working on a dramatic main staircase in the main entry, made from wood imported from the jungles of Belize, where the Coppolas own an eco-resort. The staircase was designed by Dean Tavoularis, Coppola's film set designer, implying that if the winery was not being built for a movie, it still demanded the same obsessive attention to detail.

Outside the château, men were installing computer-controlled fountains and a series of two-ton iron street lamps imported from Paris.

Led downstairs to dinner, we were seated at one long wooden rectory table set in the middle of the barrel room adjoining the vault filled with bottles of old and rare vintages of Inglenook wines.

Then began an unrestrained four-hour bacchanal, a mammoth meal of local foods prepared in the style of bistro Chez Louis, a Coppola favorite in Paris. The foie gras and duck confit repast was a marvelous showcase for several of Coppola's wines: Chardonnay, Cabernet Franc, and two older vintages of Rubicon, all followed by a rare Inglenook Cabernet. As the evening progressed it became increasingly obvious that Eleanor and Francis have a deep and abiding affection for food and wine.

So buoyed by the pleasures of the table, which by this point included Cuban cigars—and by the obvious reciprocal and collective pleasure of the guests—Francis instigated a game, a Broadway musical version of "Name That Tune." We threw out the titles. He sang, without faltering, the melodies and the lyrics. Francis knew his show tunes; he never got stumped. This little game seemed to me to say a lot about Coppola, who had found yet another way of communicating through entertaining an audience.

It wasn't until the following year, in the late spring of 1996, that I visited Niebaum-Coppola again, although I had followed the wines closely, tasting them professionally and for my own pleasure. At her invitation, I accompanied journalist Suzanne Hamlin to Rutherford for a story she was writing on Coppola the Winemaker for the *New York Times*.

Both of us, of course, knew the Coppolas through the cinema—the extraordinary films of Francis Ford Coppola had garnered him five Oscars; a documentary by Eleanor, *Hearts of Darkness* (about the making of *Apocalypse Now*, the film directed by her husband), had earned her an Emmy. We also knew the Coppolas through the media. We knew they had been married for about thirty-five years, that they had raised three grown children—Roman, Sofia, and Giancarlo, who died at age 23 in a boating accident. We knew about Coppola's plunge into bankruptcy and his struggle back to solvency, their contributions to homeless shelters and other charitable work. We knew about the resort in Belize.

What we didn't know was their unending love and respect for the land and what the land creates, and how this love and respect informs so much of their life together.

Suzanne and I met the Coppolas at the Inglenook château, and the four of us took a short walk to a magnificent bluff overlooking the recently acquired 100 acres of "front" vineyards. We sat outside in the warm sun on picnic benches. Suzanne pitched questions to the Coppolas, which were mostly answered by Francis but sometimes by Eleanor, and sometimes by both Coppolas. I occasionally threw in a few questions or comments.

The Coppolas had bought the original Niebaum home and most of the John Daniel wine estate in 1975, and in 1995 had managed to acquire the rest of Inglenook, including the Niebaum-built château. I had tasted nine vintages of Rubicon, Coppola's flagship wine, starting with the 1980 and ending with 1991. I found the wine to be, on balance, a good example of Napa Cabernet, sometimes very good but just as often disappointing.

During the course of the three-hour interview, we tasted two vintages of Rubicon, 1991 and a pre-release 1992, as well as the then-current releases of the Coppolas' Chardonnay and Cabernet Franc. As we tasted the wines, we nibbled on local mozzarella, garden tomatoes, olives, olive oil that Francis was about to label as a Francis Coppola Family product, and hunks of bread. A very pleasant afternoon.

The most compelling part of the visit was, of course, the story of the Niebaum-Coppola estate as told by a master storyteller. Francis and Eleanor, genuinely passionate about the vineyards and the winery, were able to clearly articulate their vision for the property and for their wines.

While listening to the Coppolas tell their story and the stories of Captain Niebaum and John Daniel, I realized that we were visiting the most historically significant contiguous vineyard lands in the Napa Valley, and one of the most heritage-bound vineyards in the world.

Francis spoke with great enthusiasm about the character of the wines produced on this property over the last 115 years, and the excitement of tasting the grapes in the vineyard as they ripen in the early autumn. He and Eleanor also spoke about how this historic property had been their family home for the past twenty years, how their children had attended local public schools, and how much they enjoyed being part of the Napa Valley wine community.

As we said goodbye to the Coppolas the sun began to lower over the tiny town of Rutherford. As I drove the car slowly out of Niebaum-Coppola onto Highway 29 something that Francis Coppola said kept running through my mind: "When I have to make a decision about this property, I ask myself, 'What would Gustave Niebaum do?'"

I realized that the history of Inglenook and Niebaum-Coppola was a great story, and I wanted to write that story. Unlike so many ideas that come and go, this one stuck with me.

Two months later I had written a proposal for this book and a proposal for a sabbatical leave from The Culinary Institute of America in Hyde Park, New York. I would arrive in the Napa Valley just before harvest 1997, and for almost five months I would spend just about all my days in the vineyards and the winery at Niebaum-Coppola, all my nights writing, researching, reading.

Spending such a concentrated amount of time in the Napa Valley provided a rarefied wine education. Getting to know the history of the Inglenook/Niebaum-Coppola property and the people who influenced, and continue to influence that history, made me realize that when we speak of *terroir*—the soil, the weather, the rain, all of the natural elements that affect the grapes that make the wine—that none of these elements are more important than the people who steward that soil, that land.

Walking the Niebaum-Coppola vineyards every day with winemaker Scott McLeod and vineyard historian Rafael Rodriguez, tasting the grapes as they ripen, picking the grapes, tasting the fresh juice, tasting the juice as it ferments, tasting early barrel samples, later barrel samples, tasting the trial blends, tasting the finished wines, was a magnificent gift. It was an experience that made me think about wine in a totally different way, and to savor the taste of wine in ways that are so intuitive that they often seem to challenge the words that explain them.

Through all of my work for this book—the culmination of my research—my focus is the taste of Rubicon, the ultimate expression of the *terroir* of Inglenook, of Niebaum-Coppola.

I have tasted every Rubicon that has ever been produced, beginning with the 1978 vintage. I write here about tasting Rubicon again, this time on the first day of its 1995 vintage release. I tasted the wine in the barrel long ago, and now I am pouring it from the bottle, sipping the finished wine for the first time.

To drink Rubicon is to taste nature, to inhale the bouquet of historic ground. Rubicon is a site-specific expression of the earth—an estate once whole, then fragmented, and now conjoined again. Rubicon is an evocative, purposeful, and powerful name for a wine, yet it is a humble incarnation in the pantheon of place and people: Inglenook in Rutherford; the Niebaum, Daniel, and Coppola families. To savor Rubicon is to reawaken its culture, to ignite its vision.

On this estate, a bottle of wine is not a bottle of wine. To think of Rubicon as a commodity is to ignore its viticulture and viniculture; to forget the soil; hide from the sun; not breathe the air. To experience Rubicon is to revive its ecological imperative and acknowledge the people who, through their labor and love, have defined and continue to refine the *terroir* of this estate: McIntyre, Bundschu, Deuer, Souza, Rodriguez, McLeod.

Rubicon, in all its incarnations, has always been a wine that is grown — berry by berry, vine by vine, row by row, block by block, vineyard by vineyard; this wine can never be "made." It is a wine that possesses myriad layers of complexity because the Cabernet Sauvignon, Merlot, and Cabernet Franc grapes grown in the Garden, Barn, and Gio vineyards on the Niebaum-Coppola Estate create a flavor nexus that strives for internal balance and integrity.

In the winery, the vine must not lose its voice, its grace notes. The choir of vineyards sing as one, achieving harmony and elegance.

At the moment we share Rubicon with family, friends, and food we become part of its history and culture, embarking on an interactive journey that originates with the intricacies of nature and ends with the purity of pleasure.

A Sense *of* Place

1

Rutherford Dust

In 1938, André Tchelistcheff, a thirty-seven-year-old Russian émigré and a former White Russian Army officer who fought against the Bolsheviks in the Crimea, was studying fermentation science and winemaking in Paris at the Institut Pasteur. Here he tasted a California wine — Inglenook Gewürztraminer — for the first time. And it was here that Tchelistcheff met Georges de Latour, an elegant Frenchman who owned the Beaulieu Vineyard in the very sleepy town of Rutherford in the center of the Napa Valley. Because the enologist and winemaker for the winery, Leon Bonnet, was about to retire, de Latour needed to hire someone to fill Bonnet's position. On September 15, 1938, André Tchelistcheff arrived at Beaulieu Vineyards, and the California wine industry would never be the same.

André Tchelistcheff was a short man, a little over five feet, and bore an uncanny resemblance to the actor Bela Lugosi (famous for his portrayal of Dracula and Dr. Frankenstein in Hollywood films of the 1930s). Tchelistcheff, who died in 1994 at the age of ninety-two, was never without a cigarette, even in the Beaulieu tasting rooms, to the dismay of other tasters. He was very serious, even clinical, about his work and always wore a coat and tie under a pristine white lab coat. His nickname at Beaulieu was "the Doctor."

Tchelistcheff was undoubtedly the single most influential person in the post-Prohibition Napa Valley wine industry. He introduced modern wine-making innovations such as cool fermentation and controlled malolactic fermentation, both of which led to better-balanced wines and longer lived wines. Familiar with Pasteur's work on microbes and bacteria, Tchelistcheff insisted on the most sanitary conditions in the winery.

Tchelistcheff, who was quickly made a vice president of the winery, also convinced Georges de Latour to concentrate on Cabernet Sauvignon as the flagship wine of Beaulieu. De Latour died in 1939, and in 1940 the first Beaulieu Vineyards Georges de Latour Private Reserve Rutherford Cabernet Sauvignon was released by André Tchelistcheff. For about the next fifty vintages, the Georges de Latour Private Reserve label would become not only Beaulieu's finest wine, but also the most famous and most sought-after fine wine produced in the United States. The wine is made today by BV winemaker Joel Aiken, and it is consistently a very fine wine in the intensely competitive field of Napa Valley Cabernet Sauvignon wines.

While the Inglenook/Niebaum-Coppola estate is the oldest original vineyard site in Rutherford, it was only put back together in its original 1880 land configuration in 1995. In that year, Francis Coppola, after buying the John Daniel property in 1975, bought the front vineyards and the Inglenook château from Heublein, Inc. Beaulieu Vineyards—now known as BV and owned by Heublein since 1969—was purchased by Monsieur and Madame de Latour first in 1900, as a four-acre vineyard near Rutherford, and then in 1923, as the 100-acre vineyard next door to Inglenook.

Beaulieu has produced wine every year since 1900, even during Prohibition. Georges de Latour had an exclusive and lucrative contract with the Catholic Church to produce altar wines, and Beaulieu, with a visiting rabbi in attendance, also produced kosher sacramental wines. Prohibition, which essentially ended the first wave of Napa Valley wines and wineries for so many, represented a major boom for Georges de Latour and Beaulieu.

It was André Tchelistcheff who coined the term "Rutherford Dust" to describe the unique character, the *terroir*, of Rutherford's great Cabernet vineyards, and the appearance, nose, and flavor of the finest Cabernet Sauvignon wines produced from those Rutherford estates.

Tchelistcheff, in addition to his duties at BV, became a highly regarded consultant to wineries in the Napa Valley—including Niebaum-Coppola—and indeed, around the world. He nurtured, promulgated, and witnessed the emergence of world-class Napa Valley wines, especially Rutherford Cabernet Sauvignon. His career in the Napa Valley spanned fifty-five years, and he heralded these wines through several eras, beginning with the chaotic era of post-Prohibition, and ending with the refined era of the 1980s and 1990s, when Napa Valley wines took their rightful place on the world stage.

What did Tchelistcheff mean by "Rutherford Dust," and does the term resonate when tasting Rubicon, the true descendant of the finest wines made by Gustave Niebaum and John Daniel, a singular wine from the most historically significant vineyard in Rutherford?

The town of Rutherford runs parallel to Highway 29, the dinky two-lane road that bisects the entire Napa Valley. Originally built for horse and buggy traffic when Rutherford was nothing more than a whistle stop, on a weekend during the grape harvest, today's "29" resembles a parking lot more than a highway—so concentrated is the tourist crush. As California's second- or third-largest tourist attraction (Disneyland is first, the San

Diego Zoo and the Napa Valley run neck-and-neck for second), the highway must accommodate about 25,000 cars per day on a busy weekend.

Traveling north for 3.3 miles, driving toward St. Helena and then Calistoga, Rutherford begins just a bit south of Jack and Dolores Cakebread's winery, Cakebread Cellars, established in 1973. Rutherford ends at Zinfandel Lane, where Rafael Rodriguez, vineyard historian of the Niebaum-Coppola winery, and his wife, Tila, own a home just a stone's throw from the Niebaum-Coppola estate. Rutherford is only 2 miles at its widest point, starting at Mt. St. John, which is the western border of Niebaum-Coppola and part of the Mayacamas Range, stretching to the Vaca Mountain Range due east, bordering the Silverado Trail and the Stags Leap district of the Napa Valley.

Although you get the feeling, when gazing at the Mayacamas and Vaca Mountains, that Rutherford is truly situated in a "bowl" on the Valley floor, this is really not the case. The Napa River, which runs through the center of the Valley, is 172 feet above sea level, and Rutherford vineyards extend vertically as high as 500 feet above sea level.

The mountains surrounding Rutherford are at their most majestic from daybreak to early morning. The fog that rolls into the Napa Valley, formed late at night by the cooling waters of the San Francisco Bay and San Pablo Bay, finds its epicenter in Rutherford at about 7 A.M. The Mayacamas and Vaca mountain ranges are socked in with this multilayered fog, and appear to be much higher than their actual 600 feet. As the sun burns the fog off the mountains and the vines, the most casual observer can see vineyards planted on the mountainsides.

The Napa River, after which the Valley is named, can flood easily, especially in late September and October, due to shallow banks. It has become a watershed for an increasing number, some say too many, of vineyards as the runoff from the ground finds its way to the river, which serves as a drain. Largely gone are the days, dating back to the local Wappo

Indians and stretching to the early years of the twentieth century, when the river served as an important means of transportation and a source of fish, especially trout.

Rutherford soils are generally comprised of loamy sand and gravel, specifically Yolo, Bale, and Pleasanton loams, which are somewhat richer and more fertile than, say, Bordeaux soils. The soils were formed by three alluvial fans, that is to say by three separate riverbeds, immediately following the Ice Age. The soils were formed by broken and shattered sandstone, which makes for well-drained gravel. Beneath the surface of the soil, a layering of the earth allows for water to feed the local streams and the Napa River. If we examine the pedigree of the soils, we find that the soil composition is marine sedimentary deposits and some volcanic materials.

By a quirk of nature, being that Rutherford is located in the Napa Valley's widest expanse, the area is the first part of the Valley to get late evening/morning fog that rolls off the Pacific, and the first area to get sunshine to burn off the fog. This pattern, coupled with cool nights, makes Rutherford the ideal place to grow grapes for wine. Morning fog gives the vineyards just enough moisture to disperse nutrient flow. The sun, of course, through photosynthesis in the vine, ripens the fruit. Finally, the cool nights provide a balanced measure of acidity in the grapes. Rutherford contains a wide variety of mesoclimates (often incorrectly referred to as microclimates, which really refers to the climatic differences in a canopy of vines, not an entire growing area) for such a small area, and there can sometimes be a temperature difference of 10°F between Rutherford and Oakville (the closest town to the south) and Rutherford and St. Helena (the closest town to the north).

The University of California at Davis was established by the State of California as the University Farm School in 1908, three years after the State of California purchased the 778-acre Jerome C. Davis farm. In 1922, the College of Agriculture was established. Over the years, but especially starting in the mid-1960s, UC Davis has become the leading university in the United States for its programs in agriculture, environmental sciences, fermentation science, enology, and winemaking. Davis has turned out hundreds of winemakers and enologists now working in California. The university, under the leadership of such scientists and scholars as Maynard Amerine, Albert Winkler, Harold Olmo, and Ann Noble, has undertaken important viticultural and vinicultural research, which has done much to enhance the entire wine industry in this country, but especially in California, and most especially in the Napa Valley.

UC Davis has determined that Rutherford is a Region II, meaning that it is the ideal climate for growing the red grapes Cabernet Sauvignon, Merlot, and Cabernet Franc, and the white grape Sauvignon Blanc. Using the Davis Heat Summation, which measures the high and low temperatures of a region or subregion every day from April 1 through October 31 for several years, UC Davis can determine how many degree days or heat summation units the region or subregion will contain. We look to Bordeaux as the quintessential Region II, and they grow the same grapes there as are grown in Rutherford. Carneros, at the southern tip of Napa and Sonoma and closer to the San Pablo Bay, is a cooler Region I, best for growing Chardonnay and Pinot Noir (exactly like Burgundy and the Champagne regions of France).

It is not hard to imagine why people fall in love with the scenic beauty as well as the wines of Rutherford. Rutherford covers 6,840 acres, of which 3,300 acres are under vine. Amazingly, almost unbelievably, almost 2,400 of those acres were not planted until 1989. Some of the reasons for this boom: Beginning in 1985, there was a dramatic increase in the per capita consumption of wine among Americans; from 1985 to 1991, consumption of red table wines grew about 12 percent (from 1991 to 1999, consumption has grown only 7 percent). Also, if we look at the average price Napa Valley growers received per ton of grapes in 1984 vs. 1989, the price shifts from about $768 per ton to about $1,157. If we look at 1989 prices vs. 1997 Napa harvest prices, we see a jump of 77 percent to $2,005 per ton for Cabernet Sauvignon, and a sales total of $239,737,000 for total grape tonnage sold in the Napa Valley. Rutherford Cabernet Sauvignon always commands top dollar when sold to wineries, but most of the Rutherford growers are also producers, so they sell very few tons of grapes on the open market.

Rutherford has resisted the trend in the Napa Valley toward foreign investment and multinational corporate ownership. Until 1995, the two notable exceptions were Inglenook and BV. Since Francis Coppola reunited the Inglenook estate in that year, only BV is owned by a foreign parent, Heublein, which is part of the British conglomerate, Grand Met. Rumors surrounding the possible sale of BV abound throughout the Napa Valley. Rumored asking price? $300 million. Rumored buyers? Robert Mondavi, Ernest Gallo, the Beringer Wine Estates group, which is, like Mondavi, a publicly traded company, or Jess Jackson of Kendall-Jackson wines. Francis Coppola says he is not interested.

About 2,000 Rutherford acres are planted in Cabernet Sauvignon (with grapes worth about $20 million during the 1997 harvest), with 550 acres of Merlot, and about 120 acres of Cabernet Franc. Less than 300 acres are planted to Chardonnay. Of the 137 property owners in Rutherford, 111 have vineyards; fifty-three of those vineyards are less than 25 acres, and only eight are over 100 acres. There are thirty wineries in Rutherford, and twenty-six of these wineries produce Cabernet Sauvignon wines.

If a wine is labeled as a Napa Valley Cabernet Sauvignon (e.g., Robert Mondavi) or Cabernet blend (e.g., Opus One, Insignia), consumer expectation, largely satisfied, is that the wine will be very good to excellent. When a wine is labeled as a Rutherford Cabernet Sauvignon (e.g., BV Georges de Latour Private Reserve) or blend (e.g., Rubicon), consumer expectation, again largely satisfied, leaps to the level of a connoisseur's wine—a wine of international importance.

Rutherford, an American Viticultural Area (AVA), as determined by the Bureau of Alcohol, Tobacco, and Firearms (BATF), is a subappellation of the Napa Valley AVA. The Napa Valley was granted AVA status in 1983, Rutherford in 1994. If a wine label reads "Rutherford" as the wine's controlled place of origin—what the French call *Appellation Contrôlée*—then according to BATF regulations at least 85% of the grapes that make up the wine in the bottle must have been grown and harvested in Rutherford. Legally, the remaining 15% of the grapes may come from anywhere in the United States.

Increasingly, wine consumers, connoisseurs, and collectors are paying attention to the AVA on a wine label. Ostensibly, an AVA is granted by the BATF based on the following criteria: soil, patterns of climate, elevation, rainfall, distinctive boundaries (such as mountains and rivers), historic and geographic importance, and reputation as a wine region. The BATF is petitioned by vintners and grape growers in a particular area, and hearings are held in support of, or opposition to, the petition.

There is no doubt that a recognized AVA on a label of California wine adds luster and glamour to a wine, helping to enhance its market position. Wines from such AVAs as Stags Leap, Rutherford, Oakville, and Howell Mountain in the Napa Valley, and Alexander Valley, Russian River, and Chalk Hill in Sonoma Valley have a built-in advantage in the wine marketplace. The Anderson Valley in Mendocino County is *the* appellation for fine *méthode champenoise* sparkling wines. Pinot Noir or Chardonnay from Carneros, an AVA shared by both Napa Valley and Sonoma Valley, defines the category of these wines in the United States, and is second only to Burgundy, France in worldwide reputation for fine Chardonnay and Pinot Noir.

As often as a well-known AVA helps to market a wine, lesser-known and less-revered appellations are often intentionally kept off the wine label. A reasonably priced wine grown in the land-locked and hot San Joaquin Valley or the overly productive Central Valley AVA is far more likely to use the "California" appellation, which is available to all wines produced in the state. The California AVA is also used by producers who source their grapes throughout the state, never attaining the legal 85% necessary to call a wine by a more specific appellation.

If the wine label reads "Cabernet Sauvignon," or any other varietal name (e.g., Chardonnay, Merlot), then at least 75% of the grapes that make up the wine must be Cabernet Sauvignon. The remaining 25% may be any grape or grapes. This can improve or diminish the quality of a wine, depending on how the winemaker approaches this option. Wines made from grapes like Chardonnay and Pinot Noir do not take well to blending, and the best of these will always be 100%. In Burgundy's Côte d'Or, which still represents the world standard in these wines, white wines are 100% Chardonnay, and red wines are 100% Pinot Noir. However, wines made from Cabernet Sauvignon, Merlot, or Cabernet Franc can, with judicious blending of wines made from each of these grapes, be improved. A combined 20% Merlot and Cabernet Franc can soften the tough tannins of Cabernet Sauvignon. The addition of, say, 15% Cabernet Sauvignon and 5% Cabernet Franc to a Merlot wine can give the wine more structure, balance, and complexity. In the Haut-Médoc and Graves districts of Bordeaux, where the Cabernet Sauvignon grape predominates in the wines, there is probably not one wine that doesn't benefit from at least a touch of Merlot or Cabernet Franc (and perhaps Petit Verdot and Malbec, the other two legal red grapes in Bordeaux, very little of which are grown in California).

The varietal name is coupled with the name of the AVA on a wine label. A California Cabernet Sauvignon is usually far less expensive than a Napa Valley Cabernet Sauvignon, which in turn is less expensive than a Rutherford Cabernet Sauvignon. The more specific the AVA, the more expensive the wine.

When a wine label is vintage-designated, that is, if a year appears on the label, this indicates the year in which the grapes were harvested, not the year the wine was made. The BATF requires that all vintage-dated wines contain 95 percent of grapes harvested in that vintage year. Five percent of the wine may come from older wines or younger wines. Virtually all varietal wines made in California, no matter the AVA, are vintage-dated.

If a wine is labeled "Estate-Bottled" (the rough equivalent of Bordeaux's ubiquitous "Mis en Bouteilles au Château"), this means that the same company or "estate" must have grown the grapes on land that the company owns, made the wine in its own facilities, and bottled the wine at its winery (Rubicon is always labeled as an estate-bottled wine, while Opus One, even in the years when it is totally estate-bottled, chooses not to use the phrase, so that it does not appear to be missing from the label in the vintages when Opus One is not estate-bottled). When an AVA (e.g., "Napa Valley" or "Rutherford") appears on the label of an estate-bottled wine, all of these activities must have occurred within that appellation.

If the phrase "Grown, Produced, and Bottled by" appears on the label, this might infer that although the grapes were grown in Sonoma, perhaps the wine was bottled in Napa. Alternately, it might mean that some or all of the grapes were grown on rented or leased land. "Produced and Bottled by" indicates that the wine is made from at least some purchased grapes, which, if the vineyards grow great grapes, can improve the wine. Of course, if the winery buys grapes on the bulk market, the wine may be marginal.

All of the phrases discussed above represent outward signs of quality when compared to a phrase like "Cellared by," which probably means that the winery in question bought wines in bulk, then maybe blended these wines, and bottled it under their label.

All the wines produced in the better-known AVAs are either varietal-labeled wines or proprietary wines. Wines such as Rubicon, Opus One, Insignia, and others, carry proprietary labels. So, although Rubicon may contain more than 75% Cabernet Sauvignon in every vintage (true in all vintages except 1985), legally it does not have to contain any Cab at all, or any Merlot, or any specific grape. These wines are produced and purchased based on their reputations, and like the Château-bottled wines of Bordeaux, will always carry a proprietary label. And like the wines of Bordeaux, blending of wines will vary from year to year depending on the character of the vintage. In a rainy year, more Cabernet Sauvignon to emphasize structure. In a dry year, perhaps a bit more Merlot than usual, to round out the harshness of the young Cab.*

What does Cabernet Sauvignon produced in the Rutherford appellation provide that is so unique? Why do the wine producers in Rutherford believe that the name of the place on the label is at least as important as the name of the grape? Why is it that connoisseurs of California wines would

---

* The regulations described above hold true for all states except for one: Oregon. Realizing that it cannot differentiate its wines based on quantity, because it produces less than one percent of the wine in the United States, Oregon has chosen to differentiate its wines based on quality. Oregon's regulations are as follows: AVA (e.g., "Willamette Valley"): 100 percent of the grapes must have been grown in that appellation; Varietal (e.g., Pinot Noir): 90 percent of the wine must be made from the varietal on the label, unless the wine is Cabernet Sauvignon, in which case Oregon reverts to the 75 percent federal standard. Oregon makes very little Cabernet. Finally, generic wines such as "Chablis," "Burgundy," or "Champagne" cannot be produced in Oregon. To this, the author can only add "Hallelujah."

know that if the wine is red and the label carries only one word, "Rutherford," then the wine must be Cabernet Sauvignon?

Bruce Dukes holds a master's degree in agriculture and soil science from the University of California at Davis. From 1994 to 1998, when he left California to become a winemaker in his native Western Australia, Bruce was the assistant winemaker at Niebaum-Coppola. He articulates the science and art of winemaking with elegance and ease, and his perspective on growing grapes and making wine, specifically Cabernet Sauvignon, in Rutherford merits our attention.

"I was never able to identify differences in California wines until I started tasting some of the Rutherford wines, where I can begin to identify a consistent character. The Cabernet Sauvignon of Rutherford, and especially at Niebaum-Coppola, should have very vibrant aromas, incredibly deep color, and flavors of ripe Bing cherries."

Bruce Dukes knows, as do so many other producers of fine wines, that essentially the wine is made in the vineyard, not in the winery. And Dukes also knows that the finest wines are made vine by vine, berry by berry.

"This is an incredibly beautiful area for growing Cabernet Sauvignon, which is definitely the most expressive grape in Rutherford and on the Niebaum-Coppola estate. Winemaking here is totally dictated by the property, by the grapes. Fruit from this specific property and from Rutherford in general has some of the best flavor and color characteristics that I've seen anywhere."

Just as the finest wines in the world strike a delicate balance among their flavor and aroma components, Bruce Dukes looks for that balance in the vineyard.

"Balance per vine is the absolute end of the story. We have to look at how much work we want an individual vine to do. We look at how the vine will function happily, and how that vine will ripen the amount of grapes it holds with the amount of water stored in the soil.

"The trend towards slightly closer spacing of vines is a good thing because when you get right down to it, agriculture is about farming the sunshine. If we have the sun beating down in the middle of our rows instead of on our vines, all that's happening is that the sun is heating up the soil temperatures."

Bruce Dukes reveres the wines of Rutherford for their history, their pedigree, and their sense of place.

"What we aim to do here is to make a wine that has an address and a personality that doesn't reflect California or even Napa, but says it's from Rutherford and from Niebaum-Coppola. The wine should express all of the ingredients that we've got—under the ground and above the ground—in the flavors and character of each grape."

When we taste a bottle of 1995 Rubicon from Rutherford, for which we pay $80 in a retail shop, and probably more than $150 in a restaurant, we have high expectations. First, we expect the wine to reflect the vintage, and 1995 was one of the great vintages of the twentieth century for Napa Valley Cabernet Sauvignon. Second, we expect the wine to be jam-packed with the color, bouquet, and flavors of a fine Cabernet Sauvignon. Third, but in the case of Rubicon, the most important and most elevated expectation, we expect the wine to exhibit a sense of place.

If Rubicon tastes like a good Cabernet-based wine from the Napa Valley, then the wine has missed the mark. Rubicon should demonstrate the character of the property on which it is grown, the historic Inglenook/Niebaum-Coppola estate, and it, more than any other wine produced in the center of the Napa Valley, should resonate with what André Tchelistcheff called "Rutherford Dust."

2

The Captain of the Ship

$\mathcal{I}$n 1769, Franciscan monks began the settlement of Alta California by establishing a mission in San Diego. Over the next ten years, the Franciscan friars established missions in the Santa Clara Valley and in San Francisco. The purpose of these missions was to convert the natives of the coastal valleys to Christianity, and the Franciscans were largely successful in this task.

In 1817, the San Francisco Mission decided to settle a small number of Indians at San Rafael, and the community flourished. The Indians and padres raised cattle, sheep, grains, and vegetables. They planted orchards and a vineyard, where the Franciscans, in 1821, made a small amount of wine, probably from the Criolla grape, often referred to as the Mission grape.

In 1823, Father José Altimira, with the support of Alta California Governor Luis Arguello, but in defiance of his Church superiors, established the Sonoma Mission, which he originally wanted to become the new site of the San Francisco Mission and its San Rafael *asistencia*. In a compromise with angry ecclesiastical authorities brokered by the governor,

Altimira was allowed to establish the new mission, but not as "Nuevo San Francisco." The official name of the new mission was San Francisco Solano, but it came to be known as the Sonoma Mission.

Prior to secularization of the missions by the corrupt Mexican government led by Antonio de Santa Anna, the Franciscan padres produced four types of mission wines made from the Mission/Criolla grape: a dry red; a sweet red, which nobody seemed to like very much; a dry white made from the free-run juice of the black grape, but with little skin contact; a sweet white, which was made by the addition of brandy to unfermented grape must and allowed to age for several years. This last fortified wine is the famous Angelica, which, at its best, can resemble a sweet Madeira, and still has some enthusiastic adherents in California.

A description of winemaking at the missions seems almost comical, especially when we consider that contemporary California winemaking is considered to be the most high-tech in the world. Apparently, a huge "bowl" made from sewn cowhides was hung from a platform, and the Indians dumped the grapes into this bowl. Then the Indians, wearing only loincloths, headgear to keep hair away from the grapes, and cloths wrapped around their hands to wipe away sweat, jumped into the cowhide bowl and trod the grapes. Each participant was given a stick so that he would not fall into the juice. The juice was drawn off from an opening at the bottom of the bowl and transferred to tanks. The tanks sat for a few months to allow the must to ferment. Pulp rose to the top of the tanks, and this pomace was pressed to make brandy in small copper stills. The white and red wines were consumed over a period of a few months.

The decade following the establishment of the Sonoma Mission coincided with the rise of the Mexican independence movement, which was radically anti-ecclesiastic. The mission system of Alta California was dismantled in 1833, when the Mexican government ordered the secularization of all missions. The Franciscan padres were allowed to live in the mission buildings, and became parish priests ministering to the needs of their flock, the Christian Indians. The mission lands were to be legally transferred to the Indians, but thanks to greed and corruption, this did not happen.

General Mariano Guadalupe Vallejo, on orders from the governor of Alta California, José Figueroa, dismantled and secularized the Sonoma Mission, soon to be the Sonoma Pueblo, which comprised more than 700 square miles of land, including the entire Napa Valley. Vallejo was supposed to oversee the transfer of the land to the Indians, but instead granted land to his family members, military cronies, and friends. The Santa Anna government, itself mired in corruption, did nothing to stop Vallejo.

George Calvert Yount, a native of North Carolina, first came to Alta California in the mid-1820s. He was a true American mountain man, a frontier explorer who trapped and hunted for food and trade throughout the West. He passed through the Napa Valley in 1831, taking note of its raw beauty and future possibilities. Yount traveled further south and established himself, hunting sea otters in the San Francisco Bay.

After secularization of the missions, Yount, sensing a golden opportunity, returned to the Sonoma Pueblo, and made a point of getting to know General Vallejo. He made himself useful to Vallejo by teaching the locals, who were building shelters and proper houses in the new pueblo, to make shingles from the abundant stands of Napa redwoods. Just as important to Vallejo, Yount could be counted on to kill rebellious Indians when Vallejo deemed it necessary to do away with the native people (most of whom were fellow Catholics) that he had made homeless by stealing land that was legally theirs.

Yount wanted a land grant from Vallejo, but he was not a baptized Catholic, which was necessary if he wanted to be given land in the Sonoma Pueblo. So, in 1835 Yount traveled to San Rafael, where he was baptized Jorgé Concepción. This seemed to do the trick; on March 23, 1836 Vallejo granted George C. Yount a large rancho in the Napa Valley.

Yount's land grant was Caymus Rancho, 12,000 totally undeveloped acres in the Napa Valley. The southern border of the rancho began just below what is now Yountville and extended to just about a mile below St. Helena. Yount moved to Caymus Rancho full-time in 1838, building a combination log cabin and blockhouse to protect himself from hostile Indians. He planted vegetables, acquired cattle, and hunted the ubiquitous grizzly bears, sometimes killing eight in a day. He befriended the local Napa Indians, who became his workforce, and helped to defend their tiny rancherías from marauders.

General Vallejo, who had taken over the Sonoma Mission vineyard, took about 400 cuttings from the vineyard, and some more cuttings from the San Rafael Mission, to his family's new home on the plaza of the Sonoma Pueblo. George Yount transplanted cuttings and vines from Vallejo's vineyards on his Caymus Rancho circa 1839. So it was that George Yount, frontiersman, trapper, hunter, trader, and Indian fighter, planted the first grapes in the Napa Valley.

Gustave Nybom* was born in Helsingfors (Helsinki), Finland in 1842. Finland was a quasi-independent Grand Duchy, but was really part of the Russian Empire of Czar Nicholas I. Little is known about the Nybom family, but as befits a descendant of the Vikings, young Gustave went to sea when he was about fifteen years old.

---

* As an adult living in the United States, he changed the spelling of his last name to Niebaum, because he believed it to be a more acceptable American name.

Gustave loved the sea, but realized that he needed not only the practical experience he gained while serving aboard ship, but also the technical knowledge necessary to progress through the ranks. At seventeen, he entered the Nautical Institute at Helsingfors, and graduated two years later, receiving his Master's papers in 1862. Two years later, Gustave, a hearty lad who stood six foot two, and had just turned twenty-two, received his first command. He would captain a ship through the Bering Sea to Russian America.

Russian America, today called Alaska, was claimed in the name of Czar Peter the Great in 1741 by the Danish navigator Vitus Bering, who had been commissioned by the czar to explore the regions of the North Pacific. Bering's crew returned to Russia bearing the furs of sea otters, which began the fur trade on the Aleutian Islands. The Russian fur traders enslaved the native Aleut people and practically wiped out the fur-bearing animals of the Aleutians. The Russians established the first European settlements in Russian America on Kodiak Island beginning in 1784.

In 1799, Russia chartered a trading firm, the Russian-American Company, to handle commercial and local political affairs in their colony. Until 1818, Alexander Baranof was the manager of the Company, siting its headquarters in Sitka, which he appropriated from the local Tlingit Indians. Baranof managed the Russian-American Company well enough to make money for company stockholders and at the same time make some positive inroads with many of the native Alaskans. The company sponsored Russian Orthodox priests, who traveled to Russian America to convert the natives to Christianity.

After Baranof retired, the Russian-American Company began to lose money. Russian naval commanders became the new managers of the company, and terribly mismanaged and exploited the resources of Russian America. The company held the charter to govern and develop the territory, but by any measure did a poor job, especially in developing its fur trade. This potentially profitable industry was, under the management of the Russian-American Company, a commercial and environmental disaster, with thousands of skins ruined by exploitative and excessive hunting, as well as shoddy curing practices. Due to these lack of controls, the native seal population, an important fur source, faced extinction.

The leases of the Russian-American Company expired in 1861, and partially because of its unacceptable performance, the leases were not renewed. In 1861, Russia sidestepped its navy and took direct governmental and commercial control of its outpost. Prince Maksutov was appointed by the czar to administer Russian America, remaining in his post as Imperial Governor until 1867.

Russian America remained the most remote portion of the Russian Empire when Gustave Niebaum landed in 1864 at what is now Alaska's Little Diomede Island. The Alaskan mainland was only 51 miles from Russia, and the Bering Strait—which less than 15,000 years ago was literally a land bridge between North America and Asia—had become the northernmost part of the Pacific Ocean, and was the only route to get from Siberia to Russian America.

Many considered this large landmass of frozen tundra to be a frigid wasteland, but Niebaum, a Finn who was used to the raw but livable climate, saw tremendous opportunity. Russian America was rich with natural resources: minerals, fishing grounds, and perhaps most important to the Russian market, fur-bearing animals. Niebaum realized as well that with all this natural wealth, there were opportunities to develop both trading posts and transportation facilities for this mammoth region.

Starting in 1859, the Buchanan administration began negotiations to purchase Russian America, but the distant canons of the upcoming Civil War took precedence, and the sale could not go forward. Buchanan appeared to be the ideal person to negotiate the purchase, as he was the United States ambassador to Russia during the Andrew Jackson administration, and had brokered the first commercial treaty between the United States and Russia in 1832. Buchanan was secretary of state during the James Polk presidency, and both Polk and Buchanan believed in Manifest Destiny. Buchanan pushed through statehood for the Texas territory, which led to the Mexican War. After this war, Buchanan organized the acquisition of the entire Southwest.

It was James Buchanan who brokered the final boundaries of the Oregon territory with the British in 1846. Great Britain had signed a treaty with the United States in 1818 and again in 1827 to allow citizens of both countries to settle and trade in Oregon. With the rallying cry of "Fifty Four Forty or Fight," meaning 54 degrees, 40 minutes north latitude, Buchanan established the boundaries for Oregon as part of the United States as everything south of 49 degrees. Everything north of 49 degrees belonged to England, right up to the border with Russian America.

It was not until March 30, 1867, however, that the sale was concluded, when the United States purchased Alaska for $7.4 million, roughly two cents per acre. The Crimean War had seriously weakened Russia, and with the recognition of severe mismanagement by the Russian-American Company and the Russian government, the czar was eager to sell. As we learned in elementary school, many Americans called the purchase of Alaska "Icebergia," "Seward's Folly," and "Seward's Icebox." (William H. Seward, who conducted the final negotiations with Prince Maksutov, was secretary of state in the Lincoln administration.) But Gustave Niebaum, now just twenty-five years old and living in the midst of all of this personal, social, and political upheaval, knew that the advent of modern Alaska was a significant event in world politics and commerce, and a turning point in his own life.

From 1864, when he dropped anchor in Russian America, until 1867, when the entire region became Alaska, a United States territory, Gustave

Niebaum explored the region's lands and waters, taking a special interest in the Aleutian and Pribolof Islands. He bartered for seal skins and furs with the trappers and hunters on the islands and the mainland, and began, in fits and starts, a career in trading. In 1868, a little more than three years after he first arrived, twenty-six-year-old Gustave Niebaum left Alaska, loading his furs on a cargo ship bound for California, which unlike the East Coast of the United States, still had an active finished-fur trade. As the ship sailed into the port of San Francisco, the young but confident Finnish sea captain realized how much he had learned and lived in the four years since he left Helsingfors. He had experienced the rigors and deprivations of a life at sea, and he realized that he had much more to learn and many more life experiences ahead of him. His immediate and long-term future, however, seemed bright and assured. Young Gustave Niebaum arrived in San Francisco with his own hopes and dreams, and a cargo worth more than $600,000.

San Francisco in 1868 was a prosperous, wild, exciting, and open city, simultaneously enjoying and suffering from the economic impact of the recently ended Gold Rush. It was host to pirates on its Barbary Coast, prostitutes in its famed bordellos, newly minted millionaires, and people seeking their fortunes at every level of society. Gustave Niebaum enjoyed his life in San Francisco, his adopted city, and built a fine home on Pacific Avenue. But he still had his sights set on Alaska.

Not long after selling his furs in San Francisco, Niebaum got together with some of the other traders he had met and worked with in Russian America and formed the company that would bring him an almost inestimable fortune, the Alaska Commercial Company, with offices at the corner of Sansome and Halleck. He was the youngest of the founding directors, and the only independent entrepreneur; all the others were either trading companies or financial institutions. The company, incorporated in 1869, was formed to consolidate fur-trading and natural resources operations in Alaska under a single umbrella. The United States granted exclusive rights for these industries to the Alaska Commercial Company in 1870.

Congress would not provide for an Alaskan government until 1884. Alaska was administered, in quick succession, by the War Department, the Treasury, and the Navy. Without a tax-gathering mechanism, the United States leased rights to private companies in exchange for a hefty percentage of gross profits. The Alaska Commercial Company must have represented an appealing prospect to the government, because the principals of the company were all successful, and more important, all successful in Alaska.

For twenty years the Alaska Commercial Company held leases for numerous commercial activities in Alaska. Its stewardship of the Pribolof Islands, where most of its small-scale fur trading took place, was considered exemplary, because the population of seals increased significantly on the islands, even as the company profited. The Alaska Commercial Company made lease payments over the twenty years of its stewardship of more than $9.7 million to the United States government, which was $2.5 million more than what the United States had paid for Alaska in the first place.

Niebaum was active in creating related commercial companies for Alaskan trade, using the Alaska Commercial Company as a holding company. Some of those companies were Alaska Packers, the first commercial salmon canning operation; the Northern Commercial Company, which established a chain of 100 trading posts in the Northern Territories; and the Northern Navigation Company and the Alaska Coast Company, commercial shipping lines carrying cargo across the inland and coastal water routes of Alaska on their two steamships, the Yukon Twins. These subsidiary companies as well as their parent organization were successful and tremendously profitable. During the decade of the 1870s Gustave Niebaum became a multimillionaire.

In 1873, the year that saw the appearance of San Francisco's first cable cars, thirty-one-year-old Gustave Niebaum married Susan Shingleberger, a native Californian of German descent, who was the well-bred and well-heeled next-door neighbor of one of his business partners. They were married until 1908, when Gustave Niebaum died; they never had any children.* The couple lived in San Francisco. As Gustave's fortunes increased, he decided he wanted to pursue a serious hobby, an avocation. Naturally, Captain Niebaum wanted to return to the sea, and planned to build a ship for himself and his wife so that they could take pleasure cruises whenever and wherever they wished. Susan Niebaum did not share in her husband's love of the oceans, and promptly put the kibosh on his sea-faring plans.

Gustave wanted to embrace a hobby that could be enjoyed by both husband and wife. It must have been an interesting process that led to his choice: producing one of the world's finest wines. Granted, Niebaum was by this time a sophisticated man who had traveled extensively throughout Europe, and tasted some of the world's finest wines. He spoke five languages, including French, Italian, and German, which would be enor-

---

* Five years after her husband died, Susan Niebaum's grand-niece, Suzanne, and her grand-nephew, John, Jr., were put in her care by their widowed father, John Daniel, Sr. She would raise them in the beautiful Victorian house that Gustave Niebaum built at Inglenook. The house became the home of John Daniel, Jr.'s family, and since 1975 has been owned by Francis and Eleanor Coppola, who raised their three children there.

mously helpful in a serious study of wines and winemaking. Also, he was choosing his hobby at a time when California was undergoing something of a major wine boom, mostly in and around Los Angeles, but also in Sonoma and Napa counties. In fact, Napa County boasted sixty-five wineries circa 1880 (fifty more than there were in 1980).

In 1848, James W. Marshall discovered gold at Sutter's Mill in northern California's Coloma, and so began the Gold Rush. As if to commemorate the place that the Mother Lode was found, a prospector named Stevens planted a few grapevines at Coloma in 1849. By 1860, when the thirty-first state of California was ten years old, 192,000 vines were planted throughout the golden counties of Amador, Calaveras, El Dorado, Nevada, Placer, and Tuolumne. These plantings represented a small fraction of total vines planted in the state, but at a time when southern California dominated the burgeoning wine industry, the linking of new wealth and new wine in the North was hardly coincidental.

By the 1870s, large vineyards and winemaking operations dotted the North Coast of California. Some of these companies were sidelines or even hobbies to wealthy men. Among many others who enjoyed the gentleman/ wine producer lifestyle were: Senator George Hearst, the father of William Randolph Hearst, who owned Madrone Vineyard in the Sonoma town of Glen Ellen; fellow Senator James G. Fair, for whom San Francisco's Fairmont Hotel is named, who owned the Fair Ranch, also in Sonoma; Julius P. Smith, who made his large fortune in borax mining and owned Olivina Vineyards in the Livermore Valley; Christopher Buckley, who ran the San Francisco Democratic Party with an iron hand and owned Ravenswood, just south of Livermore; and Juan Gallegos, a wealthy Costa Rican coffee plantation owner, who owned 600 acres of vineyard at his Gallegos Winery. Located in the town of Irvington, these vineyards were historically significant because the original vineyard of the Franciscan

Mission San José was part of the Gallegos property. (This winery was destroyed by the heralded San Francisco earthquake in 1906.)

Perhaps the most compelling reason to think about Gustave Niebaum's thought processes *vis à vis* his interest in wine is that the choice stands in stark contrast to his native culture and life experience. Finland and Alaska are countries of long, harsh winters and the shortest growing seasons on the planet. Grapes cannot grow in climates where the average annual temperature is less than 50°F; grapes cannot grow in Finland or Alaska. Plant life in general is restricted in both countries. Alcoholic beverages in such climates are largely based on beer, grown from grain, and spirits, made from distilled beer.

As a sophisticated and worldly man, Niebaum prized fine wine for its handmade pedigree and its civilizing character. He perceived wine as part of the good life in the civilized societies of France, Italy, and Germany. In so doing, he, consciously or unconsciously, eschewed his native culture. Just as Niebaum looked at Russian America/Alaska and saw opportunity where others saw only frozen desolation, he was now looking at the Napa Valley as a place to grow and make exquisite wines. The difference is that he was looking into this opportunity not as a twenty-two-year-old sea captain living a life of extreme hardship, but as a thirty-seven-year-old San Francisco multimillionaire.

Gustave Niebaum embraced American commerce and became rich. In choosing the Napa Valley, in choosing Inglenook above all other properties in the world available to him, Niebaum was about to embrace America as a place where the dream of producing one of the world's finest wines is possible. Once he made his decision to purchase Inglenook, the dream took shape as reality. When Gustave Niebaum came to Inglenook, he changed the face of American wine forever.

William C. Watson, secretary and cashier of the Bank of Napa, and the son-in-law of George Yount, purchased the Konig family farm, just west of Rutherford, in 1871. The farm was part of the original Caymus Rancho grant, and had been sold by Yount to Konig. Watson purchased the farm and planted a total of 70 acres, mostly planted in what were referred to as Black Malvoisie grapes, but were probably Cinsault, a grape native to the southern Rhône Valley. About 15 acres of Zinfandel had also been planted. Watson sold all of his grapes to H. W. Crabb, owner of the historic ToKalon vineyard in Oakville. (ToKalon is Greek for "most beautiful." The vineyard is controlled today by Robert Mondavi.)

It was Watson who gave the name "Inglenook" to this property. The word is an ancient Scottish idiom for "cozy nook" or "cozy corner," usually connoting the warmth of a fire and a hearth. Watson never built a winery at Inglenook and never made any wine, but he did build a resort/sanitarium that was marginally popular with San Franciscans, who would come for short restorative stays. Watson's property is that part of the Inglenook estate that borders Mt. St. John, and includes the site that its next owner, Gustave Niebaum, would choose to build his home.

In 1879, Gustave Niebaum, one of the richest men in America, decided to indulge his "hobby" of making world-class wine from a world-class wine estate. Niebaum had amassed a fortune of about $10 million, and had traveled throughout Europe and California looking for the perfect site to grow grapes and make wine. He might have easily bought one of the four Premier Grand Cru Classé estates, the "First Growths" of the Médoc in Bordeaux, classified in 1855, when Niebaum was thirteen years old. He was familiar with these *châteaux* and had tasted the wines Lafite, Latour,

Margaux, Haut-Brion.* Niebaum decided he wanted to make wines that could sit at the same table as the greatest wines of Bordeaux and the grandest of the Grands Crus of Burgundy.

Gustave Niebaum was described by his contemporaries as a man of exacting standards, a man with strong core values, but also a practical man. It must have seemed both practical and fortuitous that he found the Inglenook property about sixty miles north of San Francisco, the city in which he lived and worked as president of the immensely profitable Alaska Commercial Company.

A description of "Lovely Inglenook" from the San Francisco Examiner of April 6, 1890** helps to explain why Niebaum was so taken with the property. Clearly, the sight of Inglenook stirred both the heart and the imagination.

> Situated a few miles from the entrance to the "middle sec-
> tion" of the Napa Valley, Inglenook is a spot of indescrib-
> able loveliness in the midst of charming surroundings.

* Château Mouton-Rothschild became the fifth Premier Grand Cru Classé in 1973. Mouton-Rothschild was elevated from a "Second Growth" to a "First Growth"; the only change ever granted in the history of the Classification of 1855. Mouton-Rothschild, located in the village of Pauillac (as are Lafite-Rothschild and Latour), is owned by the heirs of Baron Philippe de Rothschild, who, in 1979, became Robert Mondavi's partner in Opus One, considered by many to be one of the Napa Valley's "First Growths."

** On this date, the *San Francisco Examiner* published a special edition on California wine. Five wineries and vineyards were featured: Gustave Niebaum's Inglenook Vineyard; Leland Stanford's Vina Vineyard; George and William Randolph Hearst's Madrone Vineyard; Charles Wetmore's Cresta Blanca Vineyard; H.W. Crabb's ToKalon Vineyard. The article on Inglenook was published without author's attribution. Sections of the lengthy article are cited throughout this chapter.

Those who only know our California scenery by a dash over the Sierras and a ride through the dust plains of the San Joaquin can have no conception of the beauties of this magnificent estate. It combines every one of the charms of scenery and climate for which the various resorts of the world are famous.

Behind the tree-bowered 'nook, from which the place takes its name, rises a chain of mountains. A tall cascade leaps down their side and goes to swell the stream that ripples through the floor of the valley. To the north, Mt. St. Helena, a few miles away, rears its head 4,000 feet above the sea and stands like a gallant sentinel overlooking an earthly paradise. East, a series of ranges of low hills covered with verdure and adorned with evergreen trees and shrubs of all kinds from the stalwart oak to the gleaming madrone add variety and charm to the landscape. South, the valley gradually narrows and then broadens out again into wide ranges of vineyards and orchards, and west, lordly redwood trees mingle with oak, laurel, and pine in decking the side of some of the grandest and loveliest canyons that ever delighted a lover of magnificent scenery. Howell Mountain and a range of hills protect Inglenook from the east winds. All the surrounding country is well wooded and well watered, and even in the driest seasons does not present that parched-up appearance, which is a common complaint about California scenery from visitors who come from more humid climates.

In 1880, Gustave Niebaum purchased Inglenook: 1,000 awe-inspiring acres in the Napa Valley, just south of the town of Rutherford, for $48,000.

The real estate records are murky, but what is clear is that Niebaum not only bought his estate from Watson, but that the land holdings of Judge Robert Hastings (78 acres) and Dorata S. Ruhlwing (440 acres) were involved in what must have been a complicated series of land sales and purchases. Whatever the machinations, on November 20, 1880 the sale was completed and recorded in the Napa County Book of Deeds. In addition to purchasing Watson/Hastings/Ruhlwing properties in 1880, Niebaum purchased smaller land parcels contiguous to the original estate in 1882, 1886, and 1887. In total, Niebaum owned close to 1,100 acres for which he paid about $60,000.*

Niebaum undertook his new, consuming hobby with the same zeal and intelligence with which he made his fortune. His commitment to his dream of making one of the world's finest wines was total, and he could not compromise his dream, since he had no expectation or desire to make any profits from his activities at Inglenook.

If Niebaum's commitment to quality seems, in historic hindsight, to be overblown, we can reference his own words to confirm that commitment. The April 6, 1890 *San Francisco Examiner* article on Inglenook opens with a quote from Niebaum addressing the issue of quality vs. profit.

> "To produce the finest wines, to equal and excel the most famous vintages of Europe, it is necessary to have the right kind of vines, grown on suitable soils, well manured, the most perfect cleanliness in handling, constant care and proper age."

---

* Over the years 1880 to 1887, Gustave Niebaum paid a total of less than $110,000 for the lands of Inglenook. A little less than one hundred years later, starting with a $2 million purchase in 1975 and culminating with a $10 million purchase in 1995, Francis and Eleanor Coppola would be able to conjoin the pieces of this original estate for a total of $12 million.

These are the rules laid down by Captain Gustave Niebaum for making wines that will compel the admiration of experts.

"No art, or trick, or machinery can make up for the absence of any of these things," says Captain Niebaum. "Grow the right kind of grapes on suitable soil, pick it when it is fully ripe, see that it is handled with scrupulous cleanliness and properly fermented, and you will have an article that has no need of the skill of the cellar master, who is a functionary that has no place in an honest vineyard save to determine in what proportion different wines should be blended."

Captain Gustave Niebaum is a San Franciscan of means who owns a magnificent vineyard and wine cellar called Inglenook, at Rutherford, in Napa County, about sixty miles from San Francisco. It is at Inglenook that he has for some years been putting into practice his knowledge of the way to make the finest of wines with such success that some time ago the Wine and Spirit Traders Society of New York, composed of the leading importers only, pronounced his wines the best California wines that had ever been shown in that state.

Captain Niebaum frankly says that he has not yet aged the best wine of which California is capable, but he has succeeded in producing wines which are superior to many of the foreign wines sold in the United States. He has performed a work of incalculable value to the state by demonstrating that California can not only produce wine, but can make a wine the equal of any in the world, and in raising the standard to a height which many either thought unattainable or so difficult of achievement that the product would not pay. He has done much to save the wine interests of the state from the threatening danger of overproduction of poor wine that could only be marketed by sophistication.

Years ago Captain Niebaum said: "I have no wish to make any money out of my vineyard by producing a large quantity of wine at a cheap or moderate price. I am going to make a California wine, if it can be made, that will be sought for by connoisseurs and will command as high a price as the famous French, German, and Spanish wines, and I am prepared to spend all the money needed to accomplish that result."

Unlike many of his contemporaries, which at this time included Charles Krug, the Beringer brothers, Jacob Schram at Schramsberg, H.W. Crabb at ToKalon, and John Benson at Far Niente, Gustave Niebaum at Inglenook was not in *the wine business*. Inglenook was Gustave Niebaum's passion, his fun, his obsession.

Having by travel, experiment, and a close study of the works of the French, Spanish, German, and Italian writers, found the most suitable vines, Captain Niebaum set to work to discover the way to make from his grapes wines that would compel the admiration of experts. This was easier for Captain Niebaum than it would be to most men, for he speaks and writes nearly all the European languages and had the time and means to travel.

As noted in the revelatory *Examiner* article, Niebaum was a serious student of grape growing and winemaking. He collected books on viticulture and enology to the point where he had one of the finest private wine-related libraries in the world. His library contained 600 books that were written in English, Italian, French, Spanish, and German, and he acquired facsimile texts of original Latin wine writings, some of them going back to the late sixteenth century. He spoke and read these languages fluently, and read the books to educate himself; enough so that in his travels to the classic wine regions of Europe he was able to ask intelligent questions of the best vignerons. It also allowed him to speak with some authority about grape growing and winemaking processes to his neighbors in the Napa Valley.

Gustave Niebaum knew that the ignoble Black Malvoisie grape would not make great wine, and he mapped out the replanting of Inglenook shortly after purchasing the property. He did not act in haste, but experimented with the varietals already under cultivation and with the rudimentary winemaking facilities on the former Hastings property. Whatever noble varieties of *vitis vinifera* he decided to plant, Niebaum knew that he had to do everything necessary to avoid phylloxera.

Phylloxera vastatrix is an almost microscopic plant louse native to the Mississippi River Valley that attacks grapevines at their roots. When cuttings of American vines traveled to Europe in the late 1850s, the plants carried the louse with them. However, because these vines were hybrids — a biological cross of *vitis vinifera* and wild grape vines — the roots of the vines had developed a strong resistance to the bug and its potentially devastating force. Still, the phylloxera survived, and when these experimental vines were planted in European soil (specifically Bordeaux, France), the phylloxera migrated to friendlier hosts, the noble *vitis vinifera*, which were planted on their own nonresistant roots. The phylloxera had a party in Europe that essentially wiped out the wine industry of France, Italy, Spain, and Germany over the course of about twenty years.

One of the less celebrated aspects of the enthusiasm for developing the wine industry in northern California immediately after the Gold Rush era was the belief that the European wine industry was finished, killed by *phylloxera vastatrix*, "the devastator." It was Louis Pasteur who, at the behest of the French government, discovered that the only way to make vines truly phylloxera-resistant is to graft vinifera vines onto native American rootstock.

As phylloxera swept Europe, few Americans realized that proud European vignerons would swallow hard and plant their vines on American rootstocks. The only other choice the Europeans had was to give up the wine trade or to grow genetically altered hybrid vines to make inferior wines.

Niebaum was possessed of a scientific mind, and understood the implications of Pasteur's research. When he ordered thousands of cuttings from Europe he knew that he would have to graft the scion to native rootstock, and Niebaum did just that. Not everyone in the Napa Valley had Niebaum's understanding of the phylloxera situation, and many of them would pay the price for their ignorance.

Phylloxera came full circle when it returned to Northern California on vines imported from France in the early 1860s. The discovery of phylloxera was not made until 1873 in Sonoma, and many of the grape growers and winemakers chose to ignore the warnings until, for some, it was too late (just as many wine producers in Napa and Sonoma would do 100 years later, when a supposedly new genotype of phylloxera was discovered in the Napa Valley).

Niebaum, who would continue to make his home in San Francisco, interspersed with frequent visits to Inglenook, realized almost immediately after purchasing the property that he would need to have a reliable person in charge of the property on a day-to-day basis. In late 1880, Niebaum hired George Mayers as superintendent of the property, so that enough land was cleared before winter to allow for planting sixty new acres in the spring of 1881. Mayers was instructed by Niebaum that the natural beauty of Inglenook was to be preserved, even at the expense of less acreage under vine. No trees on the property, especially Inglenook's magnificent oaks, were to be cut down, and any seedlings had to be protected by fences. According to the April 6, 1890 *Examiner:*

> Approaching Inglenook from the railroad, one enters the vineyard immediately by a drive through a fine avenue of almond and walnut trees, which were planted by the present owner, who is such an enthusiastic tree-planter that he laughingly says that he regards it as a religious duty to set out at least a thousand trees every year.

In February 1881, preparing for the spring planting, Gustave Niebaum purchased 1,000 Sauvignon Blanc cuttings from a nursery in San Jose. Sauvignon Blanc makes some of the best wines of the Loire Valley, such as Sancerre and Pouilly-Fumé. In Bordeaux, it is the primary grape, often blended with a touch of Sémillon, which makes the best white wines of the Graves district. With the planting of the noble Sauvignon Blanc, Niebaum's ideas about growing the finest grapes to make the finest wines took root in the soils of Inglenook.

In late 1881, Niebaum hired Hamden McIntyre as first resident general manager at Inglenook. Niebaum trusted McIntyre to carry out his philosophy; he had been an agent for the Alaska Commercial Company on the Pribilof Islands and he had viticultural experience, having worked for the Pleasant Valley Wine Company in the Hudson Valley of New York State. McIntyre, a native of Vermont, also had a background in engineering; he designed and supervised the building of Inglenook's first winery and cellars in 1883. This was a one-story building, 55 feet by 70 feet, not nearly as grand as the 1887 château winery building that still stands today, but well-built by the best stone masons.

In 1882, Inglenook produced 80,000 gallons of wine, its first vintage under Niebaum's ownership. All the wine was sold off in bulk to Lachman & Jacobi, San Francisco wine merchants. Niebaum went through all the steps of winemaking, including his fastidious approach to sanitation in the winery. He would separate the grapes from leaves, stem bits, dirt, and other organic matter before crushing. According to the April 6, 1890 *Examiner:*

> When picked, the grapes are taken in freshly scrubbed boxes to the machine room and laid on a staging. The second class bunches are put on a lateral traveler which carries them off to be made into brandy and the fine bunches are put on another traveler and sent upstairs. On their way they are subjected to a blast of air from a blower, which blows away all the dead leaves, dust and dirt, and cools the grapes. When Captain Niebaum introduced this innovation, some of his neighbors told him it was a waste of time and expense. He pointed to a huge pile of dirt and leaves that had been blown from the grapes, and asked the objectors how they would like to have it dumped in their wine vats. Before the grapes reach the crusher, they are spread out on large tables where a score of workmen armed with shears turn the bunches over and cut out every berry that is green, over-ripe or in any way damaged.

Influenced by Pasteur's pioneering work on the negative effect of microorganisms and bacteria, Niebaum insisted that everything in the winery be scrupulously clean, and spared no expense to make this a reality. All the tools in Gustave Niebaum's winery were nickel-plated for easy cleaning by the steam-cleaning apparatus he had installed in the winery.

Niebaum toured both his little winery and the later château with white gloves, and fired any workers who did not maintain his standard of hygiene.

Had he been building a palace instead of a wine cellar he could not have lavished his money more freely, and yet there was a wise economy in the expenditure of every dollar. Massive walls of hewn stone forming a building 205 feet long and 60 feet wide are much more expensive than wood or brick, but they serve to keep the temperature cool in summer and warm in winter, and the wine within can ripen perfectly. Concrete floors, made with skewed arches without a beam or truss are costly, but they give no foothold for dirt and cobwebs, and furnish no spot for the production of poisonous germs.

Every tool used for handling the grapes and working in the vats is nickel-plated. That cost money, but the tools never rust, and leave no flavor of iron in the must. Brazen door sills and brass taps, pumps and brass-bound buckets take a good deal of work to keep clean and bright, but the slightest stain shows on them and instantly reveals the fact that someone is breaking the rule of scrupulous cleanliness. The interior wainscoting and all the wooden fittings are of pine or oak, finished in oil at great expense, but such a finish shows the dirt at once and can be washed clean.

All the huge fermenting tanks are raised four feet from the floor so that the men can pass underneath them and detect the slightest leak, for leaks let drippings escape, drippings turn sour and generate the vinegar fly and cause a smell that spoils the fine flavor of the wines. Every one at

Inglenook knows that a leaky vat neglected, or a vat soured for lack of cleanliness, means instant discharge to the neglectful one, and it has been a long time since Captain Niebaum has had to dismiss a man for that cause.

Every bottle used at Inglenook is soaked in several waters, and scrubbed with a washing machine and every tub and bucket is kept as clean as the tableware in a gentleman's house, and a fierce fight is made against flies lest they should bring contaminated germs into the cellar. The slightest suspicion of acid smell in the cellar causes an investigation and the removal of the cause.

With Hamden McIntyre in place, Niebaum felt he could travel more to Europe, where he would buy vine cuttings for Inglenook, and have them shipped to California when the land was cleared of the old vines and prepared to receive new ones. He visited France, Italy, Spain, Germany, Portugal, and Hungary in search of plant material. Most of the vine cuttings for Inglenook came from France and Germany, where he had standing orders with the best nurseries; some were ordered in experimental quantities.

The *Examiner* article noted that Niebaum had to make most of the vineyard planting decisions on his own, as his grape-growing neighbors were learning on the job by trial and error.

When Captain Niebaum took possession of Inglenook he found that there was a great deal to be done before he could make a fine wine, and when he sought advice from his neighbors, he found that they differed so much about essential points that they could not help him. He also discovered that the winemakers of Europe could not get along in California. They had worked by rule of thumb all their lives, and did things because their forefathers had done them in the same way. They did not even know the names of the vines from which the famous wines were made. He soon found that winemaking here was in an experimental stage, and it was to be by practice alone that the California winemakers would achieve grand success.

By 1884, dozens of varietals were planted on Inglenook soil, many of them of European parentage, and a few from California.

The white grapes included: Palomino, from the Jerez region of Spain; Furmint, from Hungary; Burger, Franken Riesling, Moselle Riesling, Pinot Gris, and Traminer, all from Germany; Chasselas, Folle Blanche, Muscadelle, Sauvignon Blanc, and Sémillon, all from France.

The red grapes included: Cabernet Sauvignon, Cabernet Franc, Merlot, Petit Verdot, Malbec, Gamay, Pinot Noir, Carignane, Mourvèdre, Petit Sirah, all from France; Charbono, from Italy; and new plantings of Zinfandel and Crabb's Burgundy from the Napa Valley.

When it came to planting grapes, Gustave Niebaum and his neighbor in St. Helena, Charles Krug, were the only two wine growers in the Napa Valley who were familiar with the latest trials of vineyard spacing that were occurring in Europe. Niebaum had his vines planted in rows, with each row just 40 inches apart. The typical Napa Valley vineyard row was spaced eight feet apart. Studying the great vineyards of Europe, especially the vines planted in Burgundy, Niebaum discovered that when vines are planted dramatically close to each other and pruned for lower yields, each berry on the vine tasted of higher flavor concentration, complexity, and balance.

Even in 1890, Niebaum's approach to vineyard spacing was so revolutionary that it merited an explanation in a general-interest feature article. The writer for the *Examiner* observed:

> The vineyard is not laid out in the prevailing California style. Instead of a distance of seven or eight feet between the vines Captain Niebaum allows only 3 feet six inches. He does not want to get a big crop, but aims to secure a crop of fine grapes, and experience has shown that not only does close planting reduce the number of bunches and improve their flavor, but it also protects the berries from sunburn, which would otherwise follow the occasional hot spells which visit the Napa Valley when the grapes are ripening. While two vines in every seven feet is considered very close planting in this state, the most celebrated vineyards in France plant as many as 7,000 vines to the acre, or nearly four times as many, the vineyardists there being perfectly satisfied with one or two fine bunches to the vine.

Today, when people talk about how great wine is made in the vineyard, more and more they are really talking about the issue of grape yield per vine. It is amazing to realize that Niebaum used close spacing for his vines, because this is still an issue hotly debated in today's technically sophisticated, university-trained Napa Valley community of grape growers and winemakers. Some growers swear by close spacing of vines, while others swear at the concept.

Gustave Niebaum produced only estate-bottled wines at Inglenook. The grapes were grown, the wine was made, and the wine was bottled on the Inglenook estate. Niebaum did everything he could to upgrade the image of California wines, and embraced the controversial Pure Wine

Stamp program. This program, which lasted until Prohibition, was necessary because large batches of inferior wines were being shipped from Europe and being bottled as California wines. Every bottle of Inglenook wine bore the California Pure Wine Stamp, and every bottle was wired with an intricate maze so as to avoid counterfeiting the good name of Inglenook.

During the pre-Prohibition era and after, a lot of inferior California wines were sold in small wooden casks, large demijohns, and other containers. This casual practice often led to infected, "sick" wines and the further diminution of the reputation of California wines.

All of these shortcuts worked against what Gustave Niebaum was trying to achieve at Inglenook. He wanted to make the finest wines in California, and inspire others to join this quest.

The 1890 *Examiner* article made note of Niebaum's commitment to quality wines as it listed the wines available and the by-the-case retail price of each wine.

> The Inglenook wines are offered to the public in glass only, being bottled at the vineyard in California; protected by its trademark, and the Pure Wine Stamp of the State of California, which guarantees the absolute purity of the Wines; the bottles are wired, bearing the seal of the proprietor of the vineyard.

Red Wine, Table Claret, Black Label, Vintage of 1884,
per case 1 dozen quarts $3.00

Red Wine, Table Claret, Zinfandel, Vintage of 1884,
per case 1 dozen quarts $4.00

Red Wine, Extra Claret, Red Label, Médoc type, 1882,
per case 1 dozen quarts $4.50

Red Wine, Burgundy, 1883,
per case 1 dozen quarts $4.50

White Wine, Sauterne, French type, Sauvignon Vert, 1883,
per case 1 dozen quarts $4.50

White Wine, Gutedel (Chasselas), German type, 1884,
per case 1 dozen quarts $3.50

White Wine, Hock, German type, 1883,
per case 1 dozen quarts $4.50

White Wine, Burger, German type, 1883,
per case 1 dozen quarts $4.50

White Wine, Riesling, German type, 1882,
per case 1 dozen quarts $4.50
Pints per case of 24 bottles, $1.00 per case additional

It has been almost 100 years since Niebaum's stewardship of Inglenook, and the property, the winery, the culture of Inglenook, depending on that stewardship, has moved during the last century from strength to strength (Gustave Niebaum to John Daniel), strength to weakness (John Daniel to United Vintners), weakness to weakness (United Vintners to Heublein), weakness to strength (Heublein to Coppola), and finally strength to strength again (the success of Niebaum-Coppola).

Sometimes, even within the jumbled nexus of history, certain things remain constant, and the power of the Inglenook estate appears to be just such a constant, just such an imperative. While the wine business has its ups and downs, and owners of such a business come and go, it is amazing that when you come to this historical property during this new millennium, it resonates with the precise echo of a voice from the turn of the century. A section of the Inglenook article that appeared in the April 6, 1890 special edition of the *San Francisco Chronicle* describes the estate, under the subhead "A Fine View."

The unknown journalist describes Inglenook perfectly, not just for his or her then-contemporary audience, but for us as well. The writer's perceptions are ours. Inglenook, now called Niebaum-Coppola, but always Inglenook to those who know its history, is magically unchanged. The adoration with which Niebaum held Inglenook is not a thing of the past, but a thing of the present, and hopefully a thing of the future.

A Fine View

A drive of about 400 yards through the vineyard brings the visitor to the handsome stone wine cellar and machinery house. Just behind this a ridge rises dividing the vineyard. A winding road leads to the top of this ridge or knoll, from whence one can get a splendid view of the entire Napa Valley. This knoll is laid out with lanes, flower beds, shrubbery of all kinds and driveways cut out of the old solid basalt rock from which forms the backbone of the ridge. Behind the hill is a level valley, part of which is in vines and the remainder devoted to the house with its surrounding lawns, gardens, hot houses, poultry yards, stables and vegetable gardens. The vineyard proper is only 250 acres in extent; the estate includes 1,100 acres.

At the angle of the "nook" a beautiful house is now being built. The design is simple, but it is splendidly adapted for home-life in the most favored Californian climate.

"I told the architect I wanted a fine verandah, and he could do as he pleased about sticking in a few rooms," was Captain Niebaum's remark about his beautiful residence.

He managed to get his wishes carried out, for a verandah fourteen feet wide runs around three sides of the building and enables the Captain and his family to enjoy an outdoor life in all sorts of weather.

In another nook formed by the bend of the brook are the poultry houses. They are of large size and built in the most substantial manner. A wire fence about twenty-five feet high encloses the poultry yard, and inside of it the chickens, geese, ducks, turkeys, and other fowl have plenty of freedom.

For those who have walked and driven these exact paths many times, this description is as accurate today as it was more than 100 years ago. The property is magnificent, the house and its sweeping verandah spectacular, and the poultry yard still provides plenty of freedom.

Gustave Niebaum died on August 5, 1908. Because he was a well-known businessman and a multimillionaire, his obituary ran in hundreds of newspapers, but few mentioned his ownership and stewardship of Inglenook. Even the *St. Helena Star* did not mention Inglenook in its obituary of Niebaum. It seems that Niebaum was all but forgotten in the Napa Valley by the time he died, even though in the decade of the 1890s Inglenook was considered by connoisseurs, consumers, and the members of the wine industry to be America's finest wine producer.

Many have surmised that the reason Niebaum was not remembered for Inglenook was that his last surviving partner in the Alaska Commercial Company died in 1902, and Niebaum, president of the company when he died, had to turn his attention to his business affairs almost exclusively from that time. Also, for the last two years of his life, he was in failing health, and so was not a public presence in either Rutherford or San Francisco. It is certainly true that there is very little historical material to be found about Inglenook during the first decade of the twentieth century. One can only assume that as the life of Captain Gustave Niebaum went into eclipse, so too did his beloved Inglenook.

After Niebaum's death, Inglenook shut its doors and was inactive until the crush of 1911, when it was revived under a management contract entered into by Gustave Niebaum's widow, Susan. While Inglenook was up and running three years after Niebaum's death, it is inconceivable that anyone could have run Inglenook with nearly the same level of passion that Gustave Niebaum did. After all, Inglenook was Niebaum's hobby, not his business, and he never made any money from his participation in his own dream. Obviously, a management company, even with the best of intentions and excellent quality standards, looks to make a profit, even if what the management company considers a business is, at its heart, the consuming passion of one person, and has little to do with profit and loss.

Perhaps the reason that Gustave Niebaum, who, more than anyone of his age advanced the California wine industry and put it on the world stage, was forgotten at his death was exactly for the reasons that made Inglenook so successful. Niebaum was a businessman, but Inglenook was not a business. Perhaps if he had compromised a bit, and called attention to himself, and made a lot of money in the wine business, his obituary would have celebrated him as the greatest wine producer in the history of California. The irony is that if Gustave Niebaum had moved even one inch from the integrity of his dream there probably would be no Inglenook in the first place, and certainly there would be nothing to celebrate.

3

The Inheritor

*A*fter the death of Gustave Niebaum in 1908, his widow, Susan Niebaum, closed the Inglenook winery for three years. Never directly involved in the business or day-to-day operations of Inglenook, Susan Niebaum preferred to live in San Francisco. She used the estate as a weekend and summer getaway, and typical of the well-married women of the late Victorian age in America, knew nothing of business. Seeking advice from her husband's former business associates and attorneys, Susan Niebaum made some important decisions that ensured her husband's dreams for Inglenook did not die with him. From 1911 until 1919, when the 18th amendment to the Constitution—Prohibition—was ratified, fine wines were produced at Inglenook under contract with B. Arnhold & Co. of San Francisco. The vineyards and cellars were managed by Herman Lange, and the winemaking was supervised by Lafayette Stice, who lived in St. Helena.

The Arnhold Company, Lange, and Stice were unknown to Susan Niebaum, but trusting the advice of experienced people in San Francisco and Rutherford, she entered into an agreement with Arnhold to produce and market Inglenook wines. Arnhold hired Lange and Stice. The widow's decision appears to have been well reasoned, prudent, and profitable. Just as important, whether by coincidence or by design, the new winemaking team at Inglenook upheld the standards of quality set by the late Gustave Niebaum.

Stice produced 50,000 gallons of wine in 1911, and cellared wines from vintages 1908 to 1910, when the winery was dormant. Wines from the 1905 vintage and some earlier vintages sold so well that Lange and Stice decided that about 25,000 gallons of the finest wines would be set aside as reserve wines, to be released six years after vintage.

In 1912, Lange planted new vines at Inglenook on phylloxera-resistant St. George rootstock as Stice produced 80,000 gallons of wine and had 100,000 gallons of wine stored in the aging reserve. By 1915, Inglenook had once again become the California wine most sought after by serious wine lovers, winning dozens of medals in national and international competitions, including the then-important Panama Pacific International Exposition, held annually in San Francisco. At the 1915 Exposition, Inglenook won the grand prize for its Old Private Stock Sherry, the medal of honor for its Sparkling Burgundy, and gold medals for the following wines: Champagne, Haut-Sauterne, Sparkling Sauterne, Sparkling Moselle Hock, Burger, Riesling, Claret, Burgundy, Zinfandel, Chianti, Sauterne, Private Stock Port, Private Stock Angelica, Private Stock Muscat, Private Stock Madeira, and Private Stock Malaga. Inglenook won seventeen medals, more than any other California winery in the history of the international competition held each year at the exposition.

The political landscape of the country could not be ignored, however; Prohibition was on its way. In 1908, Eugene Chafin was nominated for president by the Prohibition Party, and by 1910 seven states were dry. In 1915, the town of Rutherford, home to the Inglenook estate, voted to go "dry," even as legal winemaking continued until 1919. John Daniel, Jr. was all of twelve years old.

John Daniel, Sr. was a well-known and successful engineer and contractor in San Francisco, whose young wife, the niece of Susan Niebaum, died of diphtheria in 1915. After the death of their mother, John Daniel, Jr. and his sister, Suzanne, were raised by their great-aunt, Susan Niebaum. Their father allowed the widow, who had no children of her own, to raise John and Suzanne on the draconian condition that they inherit the Inglenook estate and winery upon her death, and Susan Niebaum agreed to this condition. John and Suzanne Daniel spent their childhood and adolescence growing up in the elegant white Victorian home at the Inglenook estate.

Even during Prohibition, which ended in 1933, Susan Niebaum made sure that the vineyards at Inglenook were maintained, and sold grapes at a profit to her neighbors, the dePins family of Beaulieu Vineyards, who had a lucrative contract with the Catholic Church for the production of sacramental wines. Doubtless, John Daniel, Jr. heard stories about his great-uncle Gustave, and was groomed by Susan Niebaum (with the occasional counsel of his father, John Daniel, Sr., who continued to live in San Francisco) to take over the reins of Inglenook when, and if, Prohibition ended.

Although Prohibition made the sale of wine an illegal act, a little-remembered aspect of the Prohibition laws allowed the head of a household to produce up to 50 gallons of "fermented grape juice." This loophole in the 18th amendment was a political nod to the culture of Italian-Americans. Chances are good that utilizing the literal reading of the law and the privileges of the rich, the dining and social scene in San Francisco among the city's elite during Prohibition continued relatively unimpeded. Sharing Claret in a lovely home on Telegraph Hill was perceived as good manners, while drinking beer in a public tavern was a criminal act. As Al Capone used to say about his bootlegging activities in the suburbs of Detroit, "On the street they say I'm a criminal, but the people in Grosse Point that enjoy my product are called 'customers.'"

John Daniel, Jr. attended and graduated from Stanford, where, probably at his father's insistence, he studied engineering. John, who was passionate about aviation, wanted to become a commercial pilot in the burgeoning American airline industry, but this was considered an inappropriate and reckless career for the young man, who, by way of his family history, had a noble economic and social pedigree.

Nineteen thirty-three was a pivotal year in the life of John Daniel. That year, he graduated from Stanford with a degree in engineering but was still dreaming of flying. It was also in that year, shortly after his college graduation, that he married a beautiful, young, smart, sophisticated, outspoken woman with whom he fell deeply in love, Elizabeth Naylor. "Betty" Naylor was worldly beyond her twenty-one years, having traveled extensively throughout Europe, a great beauty who had been the object of several wealthy young men's ardor. She and John Daniel appeared to be polar opposites, which might have been the basis for their mutual attraction. He was an attractive young man who loved the outdoors, and believed in tradition. She was a rare beauty who believed that she could shape the world for her enjoyment. Their basic differences, however, coupled with Betty's Mormon faith that eschewed alcohol in any form, would lead to a rocky marriage and, inevitably, a devastating divorce.

The first year of Repeal was also 1933. So, after a honeymoon in Hawaii, John and Betty Daniel, the aspiring pilot and the aspiring socialite, returned to John's grounded and isolated childhood home at Inglenook, where they would live with John's great-aunt Susan still in residence, and where they would eventually raise their two daughters, Marcia and Robin. It appeared that John Daniel was about to enter a career that he did not pick for himself, but rather was picked for him by family tradition and the timing of American history.

Susan Niebaum died in March 1936, and as promised, left the entire Inglenook estate and winery to John and Suzanne Daniel. Suzanne had no interest in running a wine business, and left Inglenook to raise horses in Lodi, California. Since his return to Inglenook in 1933, John Daniel had become increasingly involved in the day-to-day business of the winery, tasting the wines, learning about the vineyards and the winemaking process, and perhaps most important, learning how to promote and sell the finest wines made in America. In all of these efforts he was tutored by Inglenook's general manager, Carl Bundschu, whom Susan Niebaum had hired in 1933 to lead Inglenook into its post-Prohibition years.

Carl Bundschu was fifty-six years old when he started to work at Inglenook. Born in San Francisco, he was the third generation of an esteemed Sonoma wine family; Gundlach Bundschu Winery was co-founded by his grandfather Charles Bundschu in 1858, and the family's 400-acre Rhinefarm Vineyard still produces very fine estate wines under the Gundlach Bundschu label. Carl Bundschu understood that Inglenook stood for quality, and he did everything to enhance the image of the label, including holding many open houses for members of the wine industry, the press, and well-heeled wine lovers. Bundschu also understood wine marketing. He knew that Inglenook wines had to be perceived as a rare product, and that the quality had to be in the bottle, not just in the image of the brand. For example, Bundschu did not release any Inglenook wines in 1934, a poor vintage, even though the wine-loving public had clamored for wines from the historic 1933 Repeal vintage. Bundschu hired wine-maker John Gross, who had worked at Gundlach Bundschu, and had nearly forty years' experience in the California wine industry. Gross, hired in 1934, shared Bundschu's commitment to quality and Bundschu's reverence for Niebaum's vision of Inglenook.

Bundschu figured out a way to further enhance the image of quality for the aged reserve wines made at Inglenook. In 1935, he introduced Inglenook's second label, I.V.Y. These were good wines, young wines, but not singular, great wines deserving of the Inglenook imprimatur. However, Bundschu, with the active participation of John Daniel, made sure that the I.V.Y. label carried the phrases "Inglenook Vineyard Co." and "Rutherford, Napa Co., Calif." displayed prominently on the label. These were not wines that Bundschu and Daniel were ashamed of, but affordable wines of good quality, which allowed them to raise the quality standard for Inglenook to greater heights.

Carl Bundschu was active in California's wine industry, helping to organize the Wine Institute in 1935. He was a member of the institute's first board of directors. The Wine Institute was organized after Prohibition to promote California wines, to undertake joint marketing plans, and to provide educational materials for consumers about California wines. It was, and continues to be, a lobbying group for the California wine industry, and has always featured contentious relationships among its members, especially when frequent issues of quality vs. profit have popped up. Gallo is the most powerful and most generous corporate supporter of the Wine Institute. The Robert Mondavi Winery quit the institute in 1995.

In 1936, Carl Bundschu was nearly sixty years old, and a fixture on the California premium wine scene. For the last three years he had supervised the production, marketing, and sales of Inglenook wines, and had served as a mentor for John Daniel. In July 1936, John Daniel assumed the position of general manager of Inglenook, and Carl Bundschu was appointed sales manager and director of promotion. Bundschu and Daniel worked together until the end of 1939, when Bundschu left Inglenook to become California sales manager of the Frank Schoonmaker Company, a prominent wine importer and distributor of California's best estate wines, including Inglenook, Wente, and Louis M. Martini. Frank Schoonmaker is an important historic figure in the California wine industry, because it was he who promoted the fine quality of California wines by using varietal labels only, not the then-popular generic labels, such as "Chablis" or "Burgundy."

With the departure of Carl Bundschu, Inglenook was solely in the hands of John Daniel, Jr. The Daniel Era of Inglenook, 1936–1964, is the most important in its twentieth-century history, because Daniel believed that Gustave Niebaum's dreams should and could be realized at Inglenook.

John Daniel, Jr. more than upheld the Niebaum traditions at Inglenook. He made many lasting contributions and innovations in the vineyards and the winery. He produced, starting in 1941, a vintage-dated varietal Charbono from grapes that had previously been misidentified as Barbera, then considered strictly a blending grape for generics. The Charbono became one of Inglenook's flagship wines.

Daniel believed that to ensure quality, the finest wines must truly be estate-bottled wines, meaning that the grapes were grown at Inglenook, the wine was made and aged at Inglenook, and the finished wines were bottled at Inglenook. He would not bottle inferior wines under the Inglenook label. If the wines were merely good to fine, they were sold as I.V.Y. wines, but if Daniel decided a vintage was truly poor, he would sell the wine on the bulk market. In 1945 and 1947 Inglenook did not produce a Cabernet Sauvignon for exactly this reason.

Daniel planted Chardonnay vines in a hillside vineyard at Inglenook in 1941, and was one of the first Napa wine producers to bottle varietal-designated and vintage-dated Chardonnay.

John Daniel, Jr.'s ideas and vision spread beyond his Rutherford borders. Daniel, by his commitment to quality and his no-compromise approach to producing the best possible product, literally changed the face of the American wine industry.

He was the first Napa winemaker to consistently use the name Napa Valley as an appellation on Inglenook labels, giving the wines a sense of place, and adding luster to Napa as a world-class wine region. Daniel vintage-dated the overwhelming majority of his wines as an aid to the knowledgeable consumer, and he produced varietal-designated wines, naming the primary grape on the label. This brought California wine, much of it labeled as "Burgundy" or "Chablis" or "Chianti," out of the age of generic, predictable wines, and into an age in which the grapes in the wine, coupled with their Napa Valley appellation, had a sort of individual signature.

John Daniel, Jr. was, from the beginning, the ideal steward of Niebaum's dream, right up until the end of that stewardship, when he sold Inglenook to men who would quickly end that dream, as he stood by helplessly and watched the unraveling of 85 years of history, 85 years of excellence, 85 years of Inglenook.

Robin Lail is John and Betty Daniel's daughter and the great-grandniece of Gustave and Susan Niebaum. A striking woman, about fifty-five years old, she is the strongest remaining link in the Napa Valley to Inglenook's origins. She is a woman who respects the past but lives very much in the present, with an eye to the future.

"It's a wonderful legacy being the daughter of John Daniel. For the longest time I thought it was incumbent upon me to be John Daniel or the next John Daniel, and that just wasn't possible. It took me a very long time to figure that out. I was raised as a traditionalist by John Daniel, and so the way that I live my life is informed by that tradition.

"My father was raised by his great-aunt, Susan Niebaum, and therefore he had a real Victorianism about him, and his Victorianism was not in terms of his character, or in his ability to be warm and friendly with people, but it was always evidenced in his privacy. One of his friends, in a letter, wrote, 'John Daniel, Jr. is the kind of a man with whom I can trust my most secret secret.'

"My father never spoke about family issues to anyone. Of course, I would have loved to have inherited Inglenook, but at the time you must understand that I was not raised to think like that. I was raised to believe that the property would stay, and although we never discussed it, I assumed the winery would stay. Don't ask me how it was going to stay, but it was just like: I have thumbs and I have winery; I have house and I have Inglenook.

"Our home at Inglenook was so much a part of me and it was always stated clearly clearly clearly that 'this will be yours.' But the winery wasn't part of that. Because of the family situation and the religious preferences of my mother, it just simply was never an option."

It is true that John Daniel, Jr. kept his own counsel about family and financial matters, and he was reluctant to seek publicity for his own efforts at Inglenook. But he was at the same time a tireless and important figure in the Napa wine business. John Daniel believed not only in the vines of Inglenook, but in the potential of the Napa Valley and its vignerons to produce some of the world's finest wines. He was respected by the Napa wine community and the community-at-large as a gentleman and an honest man, who wanted nothing more than to make California's best wine from the land that he had grown to love.

And John Daniel grew to love the Napa wine business, and many of the people in that business. Although Daniel was educated, wealthy, and elegant, he admired and won the admiration of his fellow Napa landowners and winemakers, some of whom had come to this country from Italy penniless and alone; others were the sons of men who made that journey. John Daniel was a friend of Louie Stralla, Charles Forni, Louis M. Martini, Fred Abruzzini, Peter Mondavi, and especially Robert Mondavi, who became his closest friend. With them, he formed the Napa Valley Vintner's Association, which is still in existence today, albeit in a much-expanded form. Louis M. Martini was the founding president of the group, and John Daniel was the founding vice president.

Daniel was also active in the Wine Institute, always arguing for quality over quantity, and for handcrafted estate wines over bulk and generic wines. He joined the board of the Wine Institute in 1941, and would remain on the board until 1969, shortly before his death. In that role, John Daniel was often referred to as the conscience of the American wine industry.

With Bundschu gone, and winemaker John Gross's retirement in 1942, George Deuer became the winemaker at Inglenook. Deuer was born in the town of Rottenburg in the then-German (now French) state of Alsace-Lorraine. He started his career as an apprentice winemaker at Christian Brothers in 1934, and then moved on to Beringer as one of several winemakers. John Gross hired Deuer in the late 1930s to work in the cellar at Inglenook, and trained him as an Inglenook winemaker. He was a rough man, and like Bundschu, enjoyed drinking whiskey a bit too much. He was a very gifted winemaker, however, and John Daniel looked to him as another mentor in his own wine education.

George Deuer and John Daniel would taste the wines together, but it was the palate of Daniel that had become the "house palate" of Inglenook. He and Deuer would taste as many as 100 wines in a day, and Daniel would identify the finest casks. He developed the "Cask Wines" program at Inglenook, starting in 1949, selling wines with their cask numbers on the label.

"Cask" was not a marketing gimmick at Inglenook; these truly were

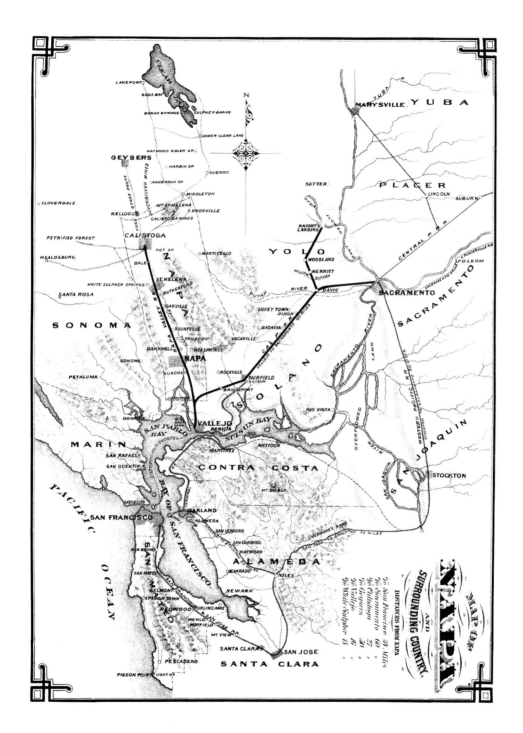

Map of Napa and surrounding counties, circa 1900. Note that Napa County,
for all its fame as the preeminent wine region in the United States,
is about half the size of neighboring Sonoma County.

The sailing ship of the Alaska Commercial Company, the St. Paul I.
The Alaska Commercial Company and its subsidiary, the Alaska Coast Company,
successfully operated commercial shipping lines carrying cargo across the inland
and coastal water routes of Alaska. The Alaska Commercial Company also
owned and operated two steamships, the Yukon Twins.

A staff photograph at one of the many outposts of the Alaska Commercial Company, circa 1880. The Company owned and operated stores throughout Alaska, in burgeoning cities as well as in the desolate frontiers of the Yukon. The stores were part of an integrated trade, cargo, fishing, and exploration company, and were highly successful, earning millions of dollars for Gustave Niebaum and his partners.

One of many woodcuts and drawings of the original Inglenook property as it appeared when owned by William C. Watson. It was Watson who purchased the property in 1871 and named it "Inglenook," a Scottish idiom for "cozy corner." He sold the 1,000-acre property to Gustave Niebaum in 1880 for $48,000.

*A photograph taken in the offices of the Alaska Commercial Company, located in a brick building
on the corner of Sansome and Halleck in San Francisco. Gustave Niebaum, one of four founding partners
in the Company, is the man seated on the far right. The man seated on the far left is Hamden McIntyre,
who was instrumental in the success of both the Alaska Commercial Company and the Inglenook winery.
As Inglenook's first resident general manager, McIntyre designed and supervised the building of the
first winery and cellars in 1883. The two other men are unidentified.*

Pickers in the Inglenook vineyards, circa 1890. While the ethnic makeup of the labor force has changed over the years, the work has not. Grapes are still picked by hand and deposited into boxes. Based on this photograph alone, the one major difference to be perceived is that the yield per vine in the vineyards is significantly less. The vine in the foreground is carrying far too much fruit to make a wine with highly concentrated flavors.

Winery workers at Inglenook, probably in the process of cleaning winemaking equipment, circa 1900.
The stone château, built by Gustave Niebaum in 1887 as a winery, is in the background.
Niebaum, familiar with the pioneering work on microbes and bacteria by Louis Pasteur,
insisted on the highest level of sanitation in the Inglenook winery.

An early Inglenook label, circa 1890. Design elements include medals
won by the winery as well as a guarantee of purity—meaning that this
was not cheap European wine bottled with a California label—
and the fact that the wines were estate-bottled.

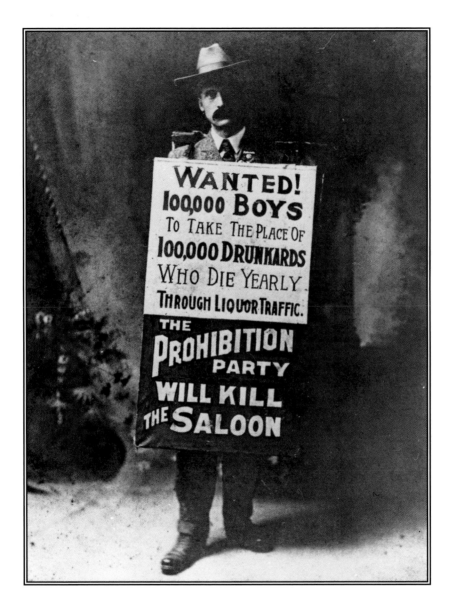

A man wearing a sandwich board encouraging voters to join and support the "drys."
In 1908, Eugene Chafin was nominated for president by the Prohibition Party,
and by 1910 seven states were dry. In 1915 the town of Rutherford, home to the
Inglenook estate, voted to go "dry," even as legal winemaking continued until 1919,
when Prohibition became the law of the land. Prohibition was repealed in 1933.

When Prohibition ended in 1933, the Inglenook château
was used for trade tastings and wine dinners. The Château
is often used for the same purposes today.

Carl Bundschu was hired by Susan Niebaum in 1933 to lead Inglenook into its post-
Prohibition years. He was the third generation of an esteemed Sonoma wine family;
Gundlach Bundschu Winery was founded in 1858 and still produces very fine estate
wines. Carl Bundschu understood that Inglenook stood for quality, and he did
everything to enhance the label's image. He supervised the production, marketing,
and sales of Inglenook wines, and served as a mentor for John Daniel, Jr.
Bundschu and Daniel worked together until the end of 1939.

Winemaker John Gross was hired by Carl Bundschu in 1934.
Gross, who had worked at Gundlach Bundschu, had nearly forty years'
experience in the California wine industry. Gross shared Bundschu's
commitment to quality and reverence for Niebaum's vision of Inglenook.

Almost from the beginning, Inglenook wines attracted critical acclaim from wine judges throughout the world.
It was not unusual for several Inglenook wines to win grand medals and gold medals in the same wine
competitions. Here, circa early 1930s, Carl Bundschu, general manager and director of sales and
marketing under Inglenook owner John Daniel, Jr., serves Inglenook wines to judges at the
Panama Pacific International Exposition, which was held annually in San Francisco.

A rare interview (on CBS radio) with John Daniel, Jr., proprietor of Inglenook, circa 1954. Daniel was the most articulate voice for the quality of Napa Valley wines, and was active in the Wine Institute. According to Rafael Rodriguez, Vineyard Historian at Niebaum-Coppola, John Daniel, Jr. "bought the very first Caterpillar tractor in Napa County."

*John Daniel, Jr. in the tasting room of the
Inglenook château circa 1958.*

A late-1950s publicity photo for the Inglenook winery. The young woman
is standing in front of a barrel with a hand-carved barrel head.
Today that barrel is on display in the Pennino tasting
room of the Niebaum-Coppola château.

the finest wines from individual casks, and the process of selection made this program wildly expensive. Rejected wines, most of which would be considered good wines by the standards of their day, were sold under the less expensive I.V.Y. label. This is not to say that the critically acclaimed Cask Wines of Inglenook were expensive; they were actually underpriced. "Pride Not Profit" was not only Inglenook's advertising slogan, it was the overall philosophy of the Inglenook vineyards and winery, and apparently it was John Daniel's personal mantra.

"My father did not make a cent at Inglenook," according to Robin Lail. "The first year the winery made any money was 1963, the year before he sold it. Remember, the Cask Cabernet was selling for $2.40. I'm sure it was a labor of love. I don't believe that you can do anything in an excellent way without it being a great love affair, and besides, my father was very Scottish in his nature, so he didn't spend a lot of money. If he weren't in love with it, he never could have done it."

There is no doubt that John Daniel enhanced not only the already estimable reputation of Inglenook, but, by his actions and his core beliefs, enhanced the overall reputation of California wines and winemaking. He did so at a time when post-Prohibition America began to slowly realize that fine wine can be a civilizing influence, and part of a fine meal. He did so during and after World War II, when many fine wines from Europe were unavailable to Americans, who cautiously looked to California to fill that gap. He did so during a time in American history when nothing seemed beyond our grasp—space flight, nuclear-powered submarines, H-bombs, Nobel Prizes in the sciences and the arts, desegregation of public schools. Americans watched television, but also read books, and about 70 percent of adult Americans drank liquor, beer, or wine.

As John Daniel approached his fifty-seventh birthday in 1963, he had to deal with the realities of fine winemaking in America. His trusted and talented winemaker, George Deuer, was about to retire from Inglenook, the winery needed a major capital infusion, and his daughters were both married. As a true Victorian gent, John Daniel would never think to bring his daughters into the family business. John Daniel began to think the unthinkable; he decided to sell Inglenook, and by 1964, he did just that.

Robin Lail remembers that "Dad offered the winery to my brother-in-law [Jim Smith, married to Robin's sister, Marcia, known as "Marky"], who was a young guy full of adrenaline, and the wine business was pretty sleepy at that time. He wanted to do something exciting, to be financially successful." John Daniel did not offer Inglenook to his other son-in-law, Robin's husband, architect John Lail, who, in fact, was interested in the winery, largely because he knew how much the tradition of Inglenook meant to his wife.

United Vintners was the production and marketing arm of a privately held grape grower's cooperative, Allied Grape Growers. The relationship between the two companies was a complex one. Allied owned United Vintners, because the co-op purchased the company from its founder, Louis Petri for $24 million in 1959.

Lou Petri was respected in the California wine industry for his ability to organize a profitable grower's co-op as well as produce and market a variety of drinkable wines at reasonable prices. Petri's empire of wineries and growers represented the fiercest competition for Ernest and Julio Gallo, who wanted to (and eventually did) become the largest wine producers in the world. When Petri heard that John Daniel, Jr. wanted to sell Inglenook he moved quickly, convinced that the Gallo brothers would buy it if they had the chance.

The irony of the Petri vs. Gallo boxing match is that all of the pugilists viewed the blood sport in exactly the same way. When the Wine Institute began an aggressive and well-funded campaign for California "premium" wines in the late 1950s, Petri and the Gallos raised their voices in protest and in unison. They objected to the idea that some California wines were "premium wines" (read wines from Napa and the North Coast) and others were not (wines made from grapes grown in the San Joaquin/Central Valley). The Wine Institute marketing campaign for "premium" wines was changed to "fine" wines, and the public relations–driven fantasy group, Premium Wine Producers of California, changed its name to the Academy of Master Winegrowers.

Petri stayed on as the chief executive of United Vintners, acquiring wineries whose vineyards became members of Allied Grape Growers. The wineries would buy grapes from Allied, and Allied was under contract to sell the grapes to the United Vintners wineries. A pretty cozy deal for all concerned.

United Vintners became the largest wine producer and marketer in California when it bought the Italian Swiss Colony winery and brand from National Distillers in 1953. United also owned the Petri, Léjon, and Mission Bell brands, and one brand with ties to Napa, Cella Vineyards. Cella, acquired by United in 1961, owned the Napa Wine Company, with a production facility in Oakville.

John Daniel sold the front 90 acres and the Inglenook château (but not his family home and the surrounding 1,500 acres) to Petri and his Allied/United conglomerate for about $1.2 million in 1964, agreeing to stay on as a paid consultant.*

---

* Five years later, Daniel's neighbor, Beaulieu Vineyards, was sold to Hartford, Connecticut-based Heublein for $8.5 million.

The president of United Vintners, handpicked by Petri, was B.C. "Larry" Solari, who was best known in the Napa Valley for buying and then liquidating the historic Larkmead winery and vineyards, rolling Larkmead into Allied/United, in the early 1950s. Solari promised and convinced John Daniel that he would maintain the reputation and image of Inglenook, and continue to make only estate-grown and estate-bottled varietal wines with the Napa Valley appellation on the label. Of course, that is not what happened.

John Daniel trusted Louis Petri to maintain the integrity, the history, the tradition of Inglenook, because Louis Petri promised him he would. John Daniel was a man of his word, and assumed that Petri was, as well. Louis Petri had acquired Inglenook as a hedge against the Gallo brothers, not because he admired what Gustave Niebaum and John Daniel had created. Once Petri and United Vintners acquired Inglenook, it was back to business as usual.

Inglenook continued to produce estate wines, but the new owners also started to produce generic jug wines under the Inglenook "Navalle" label (named for the creek that ran through the estate). The grapes for Inglenook Navalle were not sourced anywhere near the Navalle Creek, but came from Allied vineyards in the Central Valley. Inglenook was quickly becoming a brand name, not a fine-wine producer. By 1969, Inglenook had almost lost its identity; it was sold as part of the sale of 82 percent of United Vintners to Heublein, for $100 million.

When John Daniel sold Inglenook, people in the Napa Valley could talk of little else. They could not help but be surprised, and wonder why Daniel, who was the archetype for the future winemakers of the Napa Valley, and who owned the most important historic vineyard and winery in the United States, would sell his beloved property, sell his family history to people who cared little about the history and culture of Inglenook. Robin Lail has an answer to this still-burning question.

"In regards to selling the winery, it's not too surprising that it was a very big deal. Papa knew it would create a big ruckus, but that was not the point. The point was that he was extremely concerned with the viability and integrity of the brand and protection of his employees. We're discussing really simple stuff.

"He did not want this thing to drivel out as rumors that the property was for sale because he was afraid it would damage the brand severely. He was also afraid that some of his key employees might leave. So, he was very concerned about rumor and the effect it might have on the health of the business.

"He thought it would be extremely difficult to sell the property, and I don't think it ever occurred to him to offer it to his best friend, Bob Mondavi. Bob had Krug at that time, and he and his brother and his mother were busy. Bob Mondavi was absolutely his best friend in the business, absolutely. Bob thinks of my dad as one of his mentors, and Bob is my mentor in this business.

"Louis Petri, who was leading the charge at United Vintners at the time, expressed a desire to do something of real quality with the property and the brand, and that was very important to dad. It was just that simple; that's how he sold it."

Robin Lail speaks passionately when she speaks about her father and about Inglenook. Although she and her sister, Marky, did not inherit the winery from their father, he did leave them the Napanook vineyard in Yountville, about eight miles south of Rutherford, which he acquired in the 1940s. The vineyard was used to source her father's Cask Cabernet, and has been the home vineyard of the world-famous wine, Dominus, since 1982.

"My husband, John Lail, was initially responsible for me not going away from my family tradition of wine. He said, 'I want to go into the wine business,' and I remember looking at him like he had at least two heads, and asking him, 'Why do you want to do that? Why would you start over when you've already ended at Inglenook? You can't start on that plane.'" But the Lails went into the wine business.

In 1982, Robin and John Lail, Robin's sister, Marcia Smith, and her husband, Jim Smith, who live in Washington's Yakima Valley, struck an agreement with Christian Moueix, whose father, Jean-Pierre Moueix, owns Château Pétrus in Bordeaux's Pomerol district. Pétrus is currently the most expensive wine in the world. The Lails and Smiths went into partnership with the younger Moueix (who had attended UC Davis) to produce Dominus, a Cabernet Sauvignon-based wine that has received international attention.

The partnership ended in 1995, when the Lails and Smiths sold their shares in the project. Christian Moueix is now the sole proprietor of Dominus, complete with a tourist-attracting post-modern winery that opened in 1999.

Robin Lail was also a founding partner in Merryvale winery, and sold her interests in that project. She and her husband recently started a small family venture, Lail Vineyards, whose first vintage was 1995, sourcing grapes from a piece of the original Napanook, two and a half acres, contiguous to Dominus. They also own a four-acre hillside vineyard on Napa's Howell Mountain.

"We're looking at producing a maximum of 2,500 cases, but not Cabernet. We were going to go down that path, saluting my dad, but we have decided not to do that. It so happens that the Napanook property is a Merlot vineyard, so our first wine will be 56% Merlot and 44% Cabernet Sauvignon; it's a proprietary wine. It will be called Lail Vineyards J. Daniel Cuvée, and on the back label of the bottle it reads, "A Family Tradition Since 1869.""

As the proud link of this generation to John Daniel and Gustave Niebaum, what does Robin Lail think about the "New Inglenook," as presented to the world by Francis and Eleanor Coppola? Does she still dwell on what might have been if she and her family still lived in her childhood home, and still produced some of the finest wines in the Napa Valley?

"You must reach closure sometime. You have to say to yourself that this is the past, part of my wonderful, beautiful, treasured life, and just imagine how fortunate we are, as a family, to have Francis and Ellie there, and to have them be so embracing. I must say that I have an enormous, very touching bond with them, because that place is so important."

As tears well up in her eyes, Robin Lail continues to speak of the Coppolas and Inglenook. "The Napa Valley is magical and there are places within the Napa Valley that are *so magical,* and Inglenook is that. Because of the sensitivity of the Coppolas I've never had to close that book. They've been truly precious.

"Francis has gone through a lot of pathways, and he's always been so diligent about protecting his domicile, this place that he loves. He's very definitely a part of this Valley, and we'll see him become more and more a part of the Valley as time goes on. I count Francis and Ellie among my most cherished friends. It's funny how you change as you get older; it's really quite wonderful. I don't think about fate so much these days. I just like them being there. I don't know what they were thinking when they first arrived, but once the property embraced them, they were goners, done for.

"I believe that we are all just stewards of the property we have. Chief Seattle said that we are only the caretakers of the land, and that we need to make sure that the land is in pristine condition for the next generation of caretakers. Thinking about his words, that's one of the reasons I don't anguish over Inglenook; because my family had the opportunity to steward that land.

"My father felt he was the steward of Inglenook. His intention was to pass that property along, and he always treated it that way. The Coppolas are excellent stewards, because they have great passion for the land, and the magic of the land, and you just can't find a better steward than that, period."

Robin Lail rises to leave, but returns to add a coda to her musings about her connection to her beloved Inglenook.

"My daughter, Erin, married Tom Dickson, the son of a local family, two weeks ago. The wedding was at the Coppolas, and it was perfect. I mean, only the Buddha is perfect, but I'm sorry Buddha, the wedding was perfect. It was a present from the Coppolas, it was a present from the heavens, and it was the most beautiful day. Saturday, the 5th of September, 85 degrees, five in the afternoon, the light was gorgeous, the guests were wonderful, the bride and groom were so precious, so beautiful in the fullest sense of the word. Just a lovely day. I can't imagine anything more wonderful for our family."

One with the Land

The Inglenook Estate has become the Niebaum-Coppola Estate, but for Rafael Rodriguez this is a distinction without a difference. While he celebrates the memory of Gustave Niebaum, respects the integrity of John Daniel, Jr., and admires the vision of Francis and Eleanor Coppola, Rafael knows that this extraordinary land, this majestic place, can never truly be owned. The estate has its own set of imperatives that dictate its stewardship. Rafael has given himself to this historic ground and these vineyards have given him life.

At seventy-seven years old, Rafael has an abiding reverence for the history of the vineyards of Inglenook and a sense of wonder about his own place in that history. After fifty years, he continues to work this land, and the ground has become part of his being, part of his soul.

"When I walk the property, it speaks to me; I owe this property so much. I feel so grateful, so fortunate to be a link to the history of the estate. I know every square inch of the 1,400 acres, and have names for private places that nobody else knows. I planted 60 percent of the grapes in these

200 acres of vineyards. I feel that these are my vines, this is my land. If I could be buried on this property I would love it. My soul is here."

Rafael Rodriguez arrived in the Napa Valley in 1943, a lowly *bracero*, one of thousands of Mexican men who were jammed into freight trains for the long journey to California. They were to work in the fields of *el norte* during World War II, for a subsistence wage and a roof over their heads. Rafael was a city boy who had studied to work in the Mexican textile industry, but there was no work, no food, and no future for Rafael Rodriguez in México. He was brought to Napa by the World Food Administration, and was given a six-month job picking plums—the old-timers call them "prunes"—for Charlie Wagner, who years later became famous as the owner of Caymus vineyards and winery. Fifty-three years later Charlie Wagner and his son Chuck would be pictured on the cover of the *Wine Spectator*, and produce a Cabernet Sauvignon that sells for more than $100 per bottle.

Thinking about the changes in Wagner's fortunes, so emblematic of the changing fortunes of the Napa Valley, Rafael Rodriguez smiles and shakes his head.

"Seeing the success of Caymus is amazing to me. In those years Charlie Wagner's main business was prune farming. Picking prunes was one of the toughest jobs I have had in my life. I was fired after three weeks."

Rafael was afraid he would be sent to México, but "Mr. Castillo of the World Food Administration found a job for me with Salvatore Immolo. He was the only nurseryman in this area during those years making the vine cuttings that were used in the industry. At that time I had no idea what these cuttings were to be used for. It was not so long, maybe ten years after the end of Prohibition, and few people thought the wine industry had a future; it was just beginning to wake up. I worked for him for three years."

By 1947, Rafael was married to his first wife, Sally, an American citizen (who died of a chronic illness in 1953). He was beginning to feel a bit more comfortable in this land where he was still considered a temporary laborer, a Mexican who would one day have to go back to the poverty and misery of his youth. On any given day, the *bracero* program could end, and Rafael could be put on the same train that brought him to the Napa Valley, and live the rest of his life in México, with the pain of having seen the possibility of a better life, but knowing it would never be his. And then, he met Joe Souza.

"After working for Immolo, I worked in Santa Rosa for a couple of years, where I was introduced to the foreman of Inglenook, Mr. Souza. He offered me a job and housing for my wife and myself. At the time, I wasn't known in the valley, just another *bracero*, but I took the job."

John Daniel, Jr. owned Inglenook, but like his great-uncle and predecessor, Gustave Niebaum—and the Coppolas who came after him—was not trained in the wine industry. Joe Souza was Daniel's property manager, a position he held until 1971, the year after the death of John Daniel. Rafael Rodriguez knew that Daniel relied on Souza not only to do a good job with the property, but also to teach Daniel what he knew.

"John Daniel recognized Joe Souza as his mentor, and respected him highly. Souza was *the* man. He was extremely good to work with, a true gentleman. He opened the doors of the wine industry to me, which at that time was nothing short of amazing. Mexicanos were seen as a cheap source of labor and did only the most menial jobs.

"I was paid sixty-five cents an hour. Mexicanos were not allowed to prune a vine, to even get close to a vine. The top labor force was German, Italian, and then the Portuguese. At the bottom of the ladder were the Mexicans."

Joe Souza was first generation Portuguese who had worked hard as a laborer in the California orchards and vineyards, and was only one rung up from the bottom of the ladder. Looking down one rung he could see the Mexican laborers, and looking up he could see the Italians and the Germans. To hear Rafael Rodriguez tell it, Joe Souza had no desire to climb that ladder if it meant leaving a good and honest person, a hard worker, on the bottom rung.

"It was in the early 1950s. One day, to my surprise, Mr. Souza called me aside from the group and began to show me what work to do in the young vineyards. Perhaps he saw, I hope he saw, that I was capable, and had a desire to do what was necessary to care for those vines. Little by little, the vineyards began to call my attention, and I became involved with the vineyards and all that jazz required to maintain the vines. I became the man he relied on, and that created a hostile situation with the twenty-five people on the crew—the Germans, Italians, Portuguese, and especially the Mexicans, who accused me of thinking *anglo*—a little friction, you know. But I just continued doing my job, and I never kissed anybody's ass. I strongly believe that this man, Joe Souza, saw something in me.

"Joe Souza trained me to run a tractor, and that was amazing, because we Mexicans were trained to think that we should pick up the brush in the fields and pick up the garbage. When it came to equipment, to machinery, that was something else. But I, Rafael, a Mexican, learned to drive that tractor; John Daniel purchased the very first Caterpillar tractor in Napa County. I knew nothing about driving a tractor, and at first it was a disaster. Every vine I knocked down cost me two dollars. Some weeks I got no pay, but I stuck with it.

"I was humbled by and scared of that machine, but that tractor changed my life completely, completely! Once I mastered it, Joe Souza put me in charge of those vineyards. That tractor was my salvation."

In 1954, less than one year after Sally Rodriguez died, the *bracero* program ended. These Mexican peasants, virtually rented by Mexico to the United States during wartime to do back-breaking labor in the fields of *el norte*, were no longer needed, and certainly were no longer welcome. Rafael Rodriguez, who was now the father of two young children, Esther and Ralph, had been exploited for his labor during his eleven years in the United States. Yet, in the sunshine of the Napa Valley, he had begun to catch a glimmer of what his life could become. Now he was told that he must return to México, not to a life, but to a life sentence. He had to leave, but his children could stay. *They* were, after all, American citizens. Their father was a poor *bracero*, a hard worker on loan from México in the service of the *anglos*.

In the early 1950s, Woody Guthrie, the folk singer and social activist, wrote and sang the song, "Deportee," about migrant workers in the United States. He could have been singing about the plight of Rafael Rodriguez and so many other *braceros*:

> *Some of us are illegal and others not wanted,*
> *Our work contract's over, we've got to move on.*
> *600 miles to that Mexican border,*
> *They chase us like outlaws, like rustlers, like thieves.*
> *Good-bye to my Juan, good-bye Rosalita,*
> *Adios, mis amigos, you'll soon see Maria.*
> *You won't have a name when you ride the big airplane,*
> *And all they will call you will be "Deportee."*

Rafael Rodriguez's hands tremble when he remembers his fear. "I was told I had to go back to México. I was very sad, afraid that I would have to stay there. By then I was beginning to learn the American culture. I wanted to stay here, to live my life here. Discrimination in this country, compared to the discrimination and hatred I would face in México, meant absolutely nothing to me, even though I faced it, and still face it, every day of my life and in all aspects of my life. Discrimination in Napa was not as strong as it was in the cities, where it was intolerable. Here, people came to know you, and that made the discrimination more bearable."

But Joe Souza couldn't afford to discriminate; he needed Rafael, and so set about to try not to lose his right-hand man. At Souza's urging, John Daniel, Jr. wrote a letter to the American Consulate requesting that Rafael, who had no choice but to return to México, be allowed to stay—to stay at his job, to stay at his home, to stay at Inglenook, to stay in the Napa Valley, to stay in California, to stay in the United States. He had come to this country as little more than indentured, cheap labor, and now a wealthy, well-connected American, the owner of more than 1,400 acres and the most famous vineyards and winery in the United States, was writing a letter asking that Rafael Rodriguez and his children be given the opportunity to live a full and productive life. Rafael has kept this letter for forty-five years, and will keep it the rest of his life, leaving it to his children as part of his legacy.

July 20, 1955

American Consulate
Monterrey
Nuevo Leon, México

Dear Sir:

This letter is to bring up to date the certification, etc. contained in our letter of September 13, 1954 to you regarding MR. RAFAEL RODRIGUEZ SANCHEZ.

He has been employed by us on a permanent basis since November 12, 1952. His present wage is $1.15 per hour plus housing for himself and family.

We re-submit for your convenience additional information on behalf of Mr. Rodriguez's application also contained in ours of September 13, 1954:

He is a widower with two small children to support. His wife's maiden name was Sally Espinoza and she was born in La Mesa, Arizona in 1925. Her death occurred in November of 1953 while he was working here and the family was living on the premises. Their children were both born at the St. Helena Sanitarium, Napa County, California. Their names and birth dates are: Esther Rodriguez, March 23, 1948, and Ralph Rodriguez, Jr., July 8, 1949.

Mr. Rodriguez's children are now being cared for during the day by their grandparents, the parents of his deceased wife, Francisco Espinoza and Tomasa R. Espinoza, who have resided in this area for 35 years.

Unless Mr. Rodriguez's application to return to the United States permanently is granted, it will involve the breaking up of his family in addition to the loss of steady employment here and reduced earning power for their support. We sincerely hope this can be avoided.

Mr. Rodriguez is arranging to travel to Monterrey to confer with you at this time since it is the season when we can best afford to get along temporarily without his help. We sincerely hope that it may be possible to process his application at an early date so that he may return as soon as possible. We shall soon be getting into our heavy harvesting season and will need him badly at that time.

In view of the foregoing information we respectfully urge favorable action on Mr. Rodriguez's application for permission to return permanently to the United States.

We submitted a letter of identification to you from our bank, the St. Helena Branch of the Bank of America N. T. & S. A., with our letter of September 13, 1954, and are not duplicating it herewith since we understand that you have it in your files.

The original of this letter is being handed to Mr. Rodriguez for personal delivery to you.

Respectfully,

INGLENOOK RANCH
John Daniel, Jr.

Eventually, John Daniel's request was honored. After languishing in México for more than a year, Rafael Rodriguez was allowed to return to the United States. In late 1955, he became a naturalized citizen of the United States, able to live here for the rest of his life. What did all of this mean to Rafael Rodriguez? To this day, he can sum it up in five words.

"I could see my future."

Rafael Rodriguez had become part of the fabric and texture of Inglenook; a man who, even though he was still considered a third-class worker, had found a place to work and live that gave him a modicum of self-respect and job security. His was a familiar face, a familiar name to John Daniel, Jr., Joe Souza, and George Deuer, Inglenook's winemaker and general manager.

"Whenever there was a need for help in the house, the Daniel family would always call me, even though other workers had been here twenty years longer than me. Mr. and Mrs. Daniel were not happy, but they always stayed together. Joe Souza was very concerned about John Daniel's suffering in his private life."

It was known to anyone who knew the Daniel family—John, his wife, Elizabeth (known as Betty), and their two daughters, Robin and Marky—that Mrs. Daniel was very unhappy. She was a devout Mormon, and disapproved of wine and the wine business. A beautiful, vivacious, and charming woman, she was also a troubled soul. Betty Daniel "acted out" her troubles in dramatic ways, most often in the privacy of her home, but once in a while in public. She detested wine, and she was not enamored of the isolated lifestyle that she lived as the matriarch of Inglenook.

It was not only Joe Souza who was concerned about the silent sorrow of John Daniel. Everyone in the Napa Valley who knew John Daniel admired and respected him; some loved him as a role model of civility and equanimity to which they could only aspire. His best friend was Robert Mondavi, who, having been separated from his own brother and parents by an acrimonious split in the family business (Mondavi left the Charles Krug Winery to start his own), sought out John Daniel for his gracious wisdom, and looked to John Daniel's vision of Inglenook in forming his own vineyards and winery.

John Daniel was nothing if not loyal, and this extended to, first and foremost, his wife and family. It was inevitable that, if he wished to preserve that which he held sacred, he would have to leave behind that which he had built. In 1964, John Daniel sold Inglenook to Louis Petri's United Vintners/Allied Grape Growers. Petri bought a fine winery, 100 acres of historic vineyard, and the leading fine-wine brand in all of California for $1.2 million.

Rafael Rodriguez was as shocked as anyone in the Napa Valley that John Daniel had sold Inglenook, but his admiration for Daniel is undiminished.

"I cannot find words for what John Daniel did for this property and the reputation of Inglenook. He had to move slowly, because the demand for our wine was not what it is today. He never made money here, and he was not greedy; he was passionate."

John Daniel was a private man. He did not tell anyone he was selling Inglenook; not his family (his daughters, especially Robin, always thought Inglenook to be a treasured family legacy, beyond the verities of commerce), not his closest friends, and not his workers. Rafael Rodriguez was concerned about his family's future, and was worried that he would lose his job, lose his future.

Inglenook was in flux, and many people did lose their jobs or chose to leave. Joe Souza had retired and now the man in charge, winemaker George Deuer, was about to retire. Deuer was a difficult man—by all accounts a volatile personality. He was said to have a serious problem with alcohol, and he treated the Inglenook workers with contempt, especially the Mexican workers.

Rafael Rodriguez remembers that "I respected George Deuer for his knowledge and ability, but he was a strict person, really a terrible personality, and he treated me like a non-person most of the time. I was scared of that man, scared to talk to him."

Rafael had no choice but to approach Deuer, to make sure that his job in the fields was secure. There was some level of communication between the two men, but that communication flowed only one way; Deuer told Rafael what to do, and he did it.

"Deuer used to call me over to work with him. I was not allowed to ask questions, but I learned a few things about the cellar work and the basic winemaking process. So, when John Daniel sold Inglenook, I made myself talk to Deuer about my job and my future in the vineyards."

What happened in this conversation stunned Rafael Rodriguez, and as he tells it thirty-two years later, he is still stunned.

"Deuer and I got to talking about who would manage the vineyard, and he said, 'I think you can do it, Rafael. The vineyard is yours.' So, in 1965, I had my first managerial position, manager of the Inglenook vineyards; oh boy, I was excited.

"I had Joe Souza's job, and we moved into the same house that he had lived in on the property. I respected Mr. Souza until the day he died; he made my life possible. I have dedicated a tree on this property in front of the château. That tree is dedicated to John Daniel, Jr., and to Joe Souza."

Rafael Rodriguez, who had come to this country, *el norte*, as an indentured farm worker, was now, twenty-two years after his arrival in Napa, the manager of the finest vineyards in the United States. He had worked hard to attain this position, so hard that he was hospitalized more than once for literally "busting his guts."

"In the old days, when I would go to the hospital, the field worker had no benefits, no health insurance. Mr. Daniel, always a gentleman, paid my hospital bills, and made sure my family was OK. We were not close, but that is the kind of man he was."

Now, John Daniel was no longer active in running Inglenook (he was to be a consultant to United Vintners, an arrangement philosophically doomed from its start, that lasted less than two years). Joe Souza had retired, as had the irascible George Deuer. Rafael now worked for a largely faceless corporation, and he often found himself at odds with many of the decisions being made at the "new" Inglenook.

"Under United Vintners, things began to change. They would do things that John Daniel would never do on the property, even though Louis Petri promised Mr. Daniel that United Vintners would maintain the reputation of the property and the quality of the wine. They weren't interested in doing that; they just wanted the brand name of Inglenook."

John Daniel, in his role of consultant, argued vehemently against the rape of Inglenook—its rich history, its stellar vineyards, its winemaking traditions, its reputation. The United Vintners management found his attitude quaint and unrealistic, and beside the point of selling tremendous amounts of wine, which when compared to the quality wines produced on the John Daniel Inglenook Estate, could only be considered second-rate jug wine.

Daniel truly believed, at least early on, that Petri and Allied would listen to his arguments in favor of sustaining and enhancing the reputation of Inglenook. It became increasingly clear, however, that unless John Daniel was willing and able to offer suggestions as to how Allied could sell more wine he would not be taken seriously. Virtually obsessed with the image of Inglenook and the quality of its wines, Daniel might just as well have been talking to a corporate wall.

Louis Petri controlled the direction and the future of Inglenook. John Daniel retreated from his largely ceremonial consultancy, seeing his hard work and his dreams for Inglenook co-opted, usurped, but not absorbed by Petri. The concept, the idea of Inglenook as the finest wine producer in the United States, so painstakingly advanced and executed by Gustave Niebaum and enhanced by John Daniel, was dead.

It is ironic that one of the most important reasons why Allied bought Inglenook in the first place was to prevent Ernest and Julio Gallo from getting their hands on the prize vineyards and winery. But what if the Gallo brothers had bought Inglenook? Would they treat Inglenook any differently than Petri, or would they too manhandle so delicate a creature of nature?

It is difficult to speculate about a Gallo-owned Inglenook, but certain truths about the Gallo empire are inarguable. Brothers Ernest and Julio created a wine company that, while it played marketing and sales hardball for which it became infamous in the wine trade, also produced a consistent product at a very fair price. The Gallos should be credited with laying the groundwork in the 1950s and 1960s for the American wine boom of the 1980s and 1990s.

The company has always been privately owned and, with estimated revenues of more than a billion dollars, is among the top 200 private companies in the world. Ninety-year-old Ernest Gallo is still at the helm of the company that now employs three generations of Gallo family members. Gina Gallo, granddaughter of Julio (who died in a car accident in 1993) has made her mark as one of the finest winemakers in California, with the advent of Gallo-Sonoma wines. Gina's brother, Matt, is in charge of the 6,000 acres of vineyard owned by Gallo-Sonoma. Their varietal-labeled and single-vineyard wines stand in stark contrast to the traditional Gallo jug wines, still made today but slowly losing market share to California varietal-labeled wines.

Since the Gallo family is in the wine business for the long haul, and in fact has largely defined the American wine business for sixty years, it is much more likely that they would have preserved the vineyards of Inglenook. Julio Gallo, especially later in his life, was known for the care that he took in the family's smaller vineyards, and was a proponent of organic viticulture. It is unlikely that the Gallos, who have never been comfortable with opening up any of their family-owned operations to the public, would have done much with the Inglenook château, except to utilize it for its original purpose—winemaking—or for barrel storage and administration.

At the time John Daniel sold Inglenook, the fine wine boom had not taken hold in America, so it is unlikely that the Gallo family would have immediately begun to produce varietal wines from the Inglenook vineyards. In fact, the Gallos did not produce a vintage-dated varietal wine until 1978, when they made a small amount of North Coast Cabernet Sauvignon. However, once the Gallos committed to developing their Gallo-Sonoma wine program in 1977, they never turned back. Gallo-Sonoma produces more than 30,000 cases of wine annually (not even a blip on the radar screen when you consider that all the Gallo companies produce more than 60 million cases of wine). Their Sonoma vineyards are beautiful and technically remarkable; the land has been mechanically recontoured for better drainage. At their high-tech Sonoma winery in Dry Creek, the Gallos maintain the largest inventory of French and American small oak barrels—50,000 *barriques*—in the world.

If they had the will to do so, Ernest and Julio Gallo could have continued the tradition of fine wine at Inglenook, not only for themselves, but also for their children and grandchildren. One thing is certain—if Ernest and Julio Gallo had bought Inglenook in 1964, the Gallo family would, in all likelihood, still own it. They certainly would never have sold Inglenook six years after they bought it.

In 1970, Allied Vintners sold Inglenook, which had become the smallest and sexiest part of a large industrial wine conglomerate which featured Petri and Italian Swiss Colony wines as Allied's cash cows. Sales of

Inglenook wines were insignificant when compared to these jug wines. The purchaser was Heublein, a top-five giant in the spirits industry, based in Hartford, Connecticut. Heublein produced Smirnoff Vodka, the most popular vodka sold in the United States and Canada, among other highly visible products. As Rafael Rodriguez recalls, Heublein made a bad situation at Inglenook much worse.

"In 1970, Heublein bought and took over Inglenook, and things changed completely. It was a shock, but I continued to do my job. My responsibilities increased, especially after Heublein took over Beaulieu Vineyards. BV is the other historically prominent Cabernet Sauvignon property in Rutherford. I managed a total of 600 acres here, at BV, and at a property in Yountville. My salary remained the same. I had forty people working for me, all of them Mexican. Heublein began to prepare for steps they took later on, absorbing more and more acreage in the Napa Valley.

"Heublein should have never owned this property. They milked this place to nothing, and almost never made smart decisions. They told us, from their offices in Connecticut, to pull out Inglenook's Cabernet Sauvignon and plant Chenin Blanc grapes, because they thought it was more profitable. Heublein didn't care about the property. Prior to them taking over, we grew twenty-two varieties of grapes here, and they wanted us to rip out so many of them. They told us to rip out the Charbono that Niebaum planted, and to rip out the fabulous Zinfandel that grows on the property; I kind of 'forgot' to do that.

"We had always been an organic vineyard, but Heublein insisted on using the same dangerous pesticides at Inglenook that they used in the Central Valley."

Since the Gustave Niebaum era, the only sprays used in the Inglenook vineyards were small amounts of sulfur and copper, both of which are considered safe and organic. Industrial pesticides, developed during and after World War II, had long been used in California's land-locked Central

Valley, because it is the fruit and vegetable belt of the state, and depends on high yields for the livelihoods of Central Valley farmers. In addition, because most of the produce is going to be sold to consumers in supermarkets, it must look cosmetically perfect, which calls for more sprays, more insecticides.

Wine grapes do not have to look perfect, because they are not sold for eating by the public. Also, if the grower is trying to grow grapes to make a fine wine, high yields work against that idea. After a certain point, more berries per vine mean less nutrients per berry, and less flavor in the wine. In the Napa Valley, the ideal grape yield to make that fine wine is between three and four tons of grapes per acre. Fertilizers and pesticides are not needed to increase yields in Napa's vineyards. In the Central Valley, it is not uncommon for vineyards to yield from 12 to 15 tons per acre. Up until recently, to protect that kind of production, fertilizers and pesticides have been used to excess. When farm workers in the Central Valley began to develop cancers from exposure to chemicals, and growers began to lose cancer lawsuits, the use of chemicals was curtailed.

Even though the use of synthetic chemical fertilizers and insecticides was largely unnecessary in Napa Valley vineyards, growers used them because that was what farmers did during the Green Revolution of the post–World War II era. Heublein was not alone in treating their soils and vines with chemicals. The difference in their practice at Inglenook was this: Heublein, looking for the highest possible yields with the least possible hassle, converted always-organic vineyards to vineyards reliant on petrochemicals and sprays.

"Heublein was a disaster for Inglenook. The straw that broke my back was when, in 1970, the guys from Heublein decided that they wanted me to leave the Inglenook property and move to a property near Yountville. Andy Beckstoffer was in charge at that time (Beckstoffer now owns or controls more vineyard acreage than any grower in California's North

Coast counties of Napa, Sonoma, Lake, and Mendocino). I told him I didn't want that assignment and that I deserved more money for my job, and he didn't deal with either issue, so I gave my notice. I couldn't believe it myself. My God, I quit Inglenook."

Rafael Rodriguez could not have foreseen or planned the next five years of his life, from late 1970 to early 1976. He was about to embark on another journey, another struggle, leaving his beloved Inglenook behind. Rafael was compelled to present himself to the larger culture, to re-invent himself, not as a vineyard worker, but as a respected citizen holding elective office, as a controversial labor organizer, and as a teacher.

"In the 1960s, Mexicans began to realize that living conditions in the Napa Valley, even though they were meager, were better than in other agricultural areas. They began, as I did when I was a young man, to think about staying in Napa, to send their children to our schools, to make a better life, if not for them, then for their children."

With some prodding from the superintendent of the St. Helena school district, who realized that with 10 percent Mexican children in the schools, the complexion of the student population was changing, Rafael agreed to campaign for a seat on the St. Helena school board. He was elected to the office, and later became board president.

Rafael Rodriguez, the Mexican immigrant who barely escaped being sent back to live his life in the slums of México City, had become an active and responsible American citizen. Even with these enormous changes in his life, Rafael never forgot who he was; he was not an assimilated Mexican, a fake *anglo*. He knew, better than anyone, that his job on the school board was providing a voice for the children of farm and vineyard workers.

At the same time that Rafael was representing the interests of Mexican and Mexican-American children, he also wanted to represent what he considered to be the best interests of their parents, who, like Rafael himself, toiled on the farms and in the vineyards of the Napa Valley.

By the late 1960s, the United Farm Workers, headed by the charismatic and strategically brilliant Cesar Chavez, had begun to penetrate the Napa Valley. Heublein, Christian Brothers, Krug, and Trefethen had all signed contracts with Chavez, largely out of fear that if they did not sign, he would cripple the fine wine industry of the North Coast, as he had crippled the vineyards of the Central Valley by a wildly successful UFW-led boycott.

Chavez and the UFW were not nearly as powerful on the North Coast, especially in the Napa Valley, as they were in the Central Valley, where near–slave labor conditions were the norm. The smaller North Coast wine producers resisted signing with the UFW. They realized that to be successful against Chavez, they, like the workers, had to organize. They formed the Growers Foundation, and in a show of enlightened self-interest, offered a better deal than the UFW to their workers. Issues on the table were pensions, insurance, vacations, transferable benefits from one job to the next, and an increase in the minimum wage. On every point, the Growers Foundation came up with better numbers than the UFW.

Chavez spoke for the worker and in the language of the worker, making his demands to the *gringos* in English, and passionately explaining them to the farm workers in Spanish. Why would the workers listen to the owners? Their message was good, but they had nobody to articulate it.

Rafael respected Cesar Chavez, and thought he did important work in the Central Valley, but did not support Chavez and the UFW in Napa. Heublein, and by extension, Inglenook, had signed with the Farm Workers during Rafael's tenure as vineyard manager. Although Rafael believed the workers needed representation, he did not like the impact that the union had on the workers, and he believed that the philosophical under-pinnings of the UFW were undesirable, even subversive to the interests of the workers.

"The Chavez ideology was influenced by communism, and some of the Mexican workers groups decided they didn't want that. Socialism in México led to no progress. Heublein and others signed with the UFW, but through the years, some of the Mexican union workers began to move away from the union."

Looking away, looking at the vineyards in the distance, Rafael adds, "You have no idea how the union movement divided friendships and families."

The Growers Foundation came to Rafael Rodriguez, who, by 1970, was known to many of its members as the vineyard manager at Inglenook and as a school board member. They knew that Rafael was uniquely qual-ified to sell their plan to the workers. Rafael questioned them closely; he had attended too many school board meetings with people who mouthed politically correct rhetoric when it was politically expedient. He wanted to make sure that the owners were sincere in their proposal, and would not go back on their word to him and to the workers.

The Growers Foundation offered Rafael triple the salary he was mak-ing at Inglenook, which was, of course, tempting. It was clear that if he was to throw his lot in with these owners, his good name would be maligned in his own community, and he might lose the respect of his friends and coworkers. Rafael knew, of course, that to go up against Chavez and the UFW was a life-threatening activity, and his life was worth more than money. He realized that even though he was miserable at Inglenook, and he was staring at a $25,000 salary—more money than he could ever imagine earning—the only reason to take the job as field coor-dinator with the Growers Foundation was that he believed in their offer to the workers, and believed that through this work he could make a differ-ence in the lives of those workers.

"I got involved with the Growers Foundation because it was a better alternative to Chavez and the UFW. The union could not compete with us. We had a good program: twenty cents more per hour than the UFW contract, employer-paid worker's health benefits, paid vacation, pension paid by the employer. Still, it put me in a very difficult position.

"I was in danger because Mexican workers believed that if you were not a *chavista* you were not a Mexican. So when they saw me holding a different flag, they called me so many names that I had to put them out of my mind and out of my heart. They thought I was trying to be *anglo,* and they denigrated my name and the honor of my family. It hurt so much because it came from Mexicans. They could not understand that they had a free choice to either choose the union plan or choose the Foundation plan."

In the end, the Growers Foundation came to represent about 1,500 vineyard workers. Rafael was let down, not only by his experience with his own people, but with some of the growers. He had to demand that they stick to their promises. He had promised many things to the workers on behalf of these *anglos,* and had extended his hand in friendship and the hope of solidarity with fellow workers. He could not tolerate any duplicity by any of the growers; he had far less patience for them than he did for Chavez.

"I left the Growers Foundation after four years. I quit when, after the threat of the UFW subsided, the benefits changed. No pension plan, employee contributions for insurance, and bad housing was still not addressed. I could not represent the Growers by lying to the workers. I had to quit."

Starting in 1972, while still working for the Growers Foundation, Rafael Rodriguez became a part-time college teacher, working as an adjunct instructor for Napa College.

"The college began to hold seminars on pruning and budding, and for all the practices in the vineyards. This was an amazing opportunity for me to implement a field budding class, teaching in the vineyard how to graft a vinifera vine to the appropriate rootstock.

"When I supervised Inglenook and Beaulieu we did field budding all the time. I believe in it 100 percent. If there is a problem with the graft you can take steps right away to correct it on the spot. When you accept grafts done in the nurseries, you can receive thousands of cuttings, some of them in poor condition. Also, some of the new rootstock has not been properly identified and certified; we are still learning about its behavior.

"I used to travel to three different counties—Napa, Sonoma, and Mendocino—to offer classes in pruning and budding, in Spanish and English. I did that for three years until late 1975, but I felt a little defeated in this program.

"The program was geared to train Mexicanos, to give them important skills, but only *anglos* would show up for the classes. At the time there was a cultural block in education; that's changing now.

"I was shocked to be teaching winemakers and vineyard managers, but these students respected me. They didn't think of me as a Mexican, but as a teacher. Later, I would read articles about the work a winemaker or vineyard manager was doing, and the person being interviewed would say that he was doing this work because of what I taught him. That really pleased me."

Early 1976 found Rafael Rodriguez, now fifty-five years old, husband and father, homeowner, vineyard manager, reluctant labor organizer, and newly minted educator, looking for a new challenge. How could he have known that his new challenge would be to resurrect the old Inglenook?

"One day I came home and there were two people sitting in my living room. The one I knew was Michael Bernstein, who was then the owner of Mt. Veeder Winery. He had been a student of mine in the Napa College budding seminar. I had no idea who the other man was. Mr. Bernstein introduced me to Francis Ford Coppola. I had never heard of him, and had no idea who he was or why he was in my house.

"We were introduced and Francis said that he had just bought the old John Daniel property, that part of Inglenook that included the Niebaum house and vineyards, and hundreds of acres bordering Mt. St. John. Anyway, he bought all of Inglenook that didn't belong to Heublein—the front 100 acres and buildings facing Highway 29.

"And then Francis asked me if I would come back to the property to manage it. I just couldn't believe it. He said he would double my salary, and wanted me to manage the entire property, not just the vineyards. After thinking about it for a while, I resigned officially from the Growers Foundation, and agreed to return to the Niebaum Estate.

"When I returned here in 1976, my salary was one of the highest paid in the Napa Valley. I went from $25,000 to $40,000. Francis said, 'You work for me, you will be part of an ownership program, you will have benefits, vacation, retirement.' In this industry, the workers have Chavez to thank for introducing benefits, but Coppola did so because that is what he always wanted to do.

"The dreams that Francis had to begin with were beautiful. He wanted to make a nature preserve here and revitalize what John Daniel had. His first words to me were 'Rafael, I want this place to be kept as if John Daniel still owned it, but better.'

"The Coppolas really care about the property. Both of them, especially Eleanor's vision for the property is in keeping with its heritage. People say Francis is a dreamer, but I think he is a very persistent person—'If I don't get it today, I will get it tomorrow, or the next day.' I identify with his persistence; I am like that, but on a different level. I want you to find out who I am without me telling you. It doesn't pay for me to explain myself. If you do not appreciate what I do for you today, that's OK, but tomorrow you will."

Rafael returned to this property that was such a big part of his life to work for someone whom he respected, and someone he believed in. He felt that the spirit of John Daniel, Jr., so long ignored on the adjacent Inglenook property owned by Heublein, was alive and well on the Coppola property, which Francis decided to call Niebaum-Coppola. He did so to credit Inglenook's founder, and to model the estate on the finest Bordeaux châteaux, which, like Lafite-Rothschild, often combined the names of previous and current owners to establish an historical identity.

While the Coppolas understood and celebrated Inglenook's lineage and the importance of its history, the first fifteen years of their ownership, from 1975 to 1990, was financially rocky. Coppola's *Apocalypse Now* bankruptcy, while often exaggerated in the media and on the wagging tongues of the Napa Valley, had a profound effect on the vineyards.

"For fifteen years there was no money. I had only one tractor and things were tough for my workers, for the Coppolas, for me. We were afraid Francis would have to sell the place, but he stuck with it. He never even mentioned selling. He has a true feeling for this place. What I admire about Francis is that he comes to a stoplight and, Jesus Christ, it means nothing to him. He doesn't stop and he succeeds."

No one would argue that Francis Coppola is not, at base, a businessman. He is an artist, and he is not particularly "practical" in the conventional sense of that word. On the other hand, when he faced bankruptcy, he did what he had to do, which was to become a director for hire, in order to hold on to his dreams for the Niebaum-Coppola Estate. And ironically, *Apocalypse Now*, which initially was the cause of his bankruptcy, has become profitable, to date grossing more than $125 million, and may well turn out to be his most personally profitable film. He had to buy the film to secure its completion and distribution, which put him more than $20 million in personal debt. So, he owns, outright, the definitive film about the Vietnam War; some would say it is *the* film about all wars.

As Francis worked his way out of bankruptcy, life in the vineyards of Niebaum-Coppola began to ease up and become more productive. And Rafael Rodriguez was the first to notice the difference.

"Things are good here now, we have money, and Francis is more involved. Today, we have five tractors, and we can do our jobs well."

One of the dreams that Francis and Eleanor Coppola held on to, even during their darkest financial days, was to reunite the original Inglenook estate, which meant buying the 100 acres fronting Highway 29, the château, and the barrel storage building—a monstrosity built by Heublein that blocked the view of the château from 29, and which Coppola, fresh from the pyrotechnics of *Apocalypse Now,* often mumbled about blowing up.

In 1975, the Coppolas bought the original Niebaum home and property with profits from *Godfather II* for about $2.5 million. Twenty years later, Francis and Eleanor Coppola, who had faced almost insurmountable debts, but never gave up their dreams, purchased the remaining property and buildings from Heublein for more than $10 million with profits from *Bram Stoker's Dracula.* This was an historic moment for the Napa Valley and for the American wine industry.

Rafael Rodriguez, who had come to know and love this property under the stewardship of John Daniel, and had become alienated from it under the ownership of Heublein, rejoiced.

"When Francis bought the other piece of Inglenook, it was glorious. Just to know that the whole property came together, which was something I never dreamed I would see again in my lifetime, was exciting.

Francis was so excited, it was contagious. This is the only historic vineyard in the Napa Valley that is in one piece.

"There are many beautiful places in the world that are highly recognized for what they have done throughout the history of winemaking. Inglenook has to be recognized as one of them. The legacy of Gustave Niebaum, John Daniel, Jr., and Francis Ford Coppola, the spirit of Inglenook, and its inheritor, Niebaum-Coppola, will continue with excellence. I am sorry that I am too old to continue at full throttle, but the property rejuvenates me. John Daniel would be so happy to see that this property has come back to be in one piece, that its character, its *terroir,* and its legacy have been preserved. It is a dream come true."

And what of Rafael Rodriguez's dreams?

"One of my dreams was to visit Europe to see the vineyards. I had tasted the wines of the Domaine Romanée Conti, and wonder why they charge $800, and why we charge $30 for pinot noir. I went to France, sent by Francis, and was treated as if I was the owner of Inglenook. Madame Lalou (Lalou Bize-Leroy, the former proprietor of the DRC) was asking me my opinion about her wines, which were distributed by Wilson-Daniels, the importer in St. Helena, good friends of mine that my daughter worked for. When I saw the Domaine Romanée Conti, I just couldn't believe it.

"Then I went to visit Dr. Biondi-Santi at his historic vineyards in Brunello di Montalcino. Again, I couldn't believe I was there. Later, we introduced our wines together in San Francisco, when we showed the '79 Rubicon.

"The company sent me to Spain for three weeks, and Rioja was the highlight. In Portugal, Oporto was so impressive, because I thought of

Joe Souza, who was Portuguese. Seeing where he came from, that is when I understood what kind of person he was, and my heart was opened to the people of the Douro. All these things I owe to this property."

And the Coppolas must believe that the property owes a lot to Rafael. Every year the Coppolas hold a harvest party for family, friends, and all the Niebaum-Coppola workers and their families. At the 1997 harvest party, Francis presented a clear glass plaque, adorned with a portrait of Rafael, and announced that one of the vineyards on the property would be named for the Niebaum-Coppola Vineyard historian, Rafael Rodriguez.

"It was a surprise, an honor, a dream come true that a vineyard on this property would be named after me. I think it will be the vineyard on the left as you enter the property, which was planted just prior to Francis taking ownership of that property.

"As the historian of the estate, I give private tours of Niebaum-Coppola to wine professionals, business people, politicians, artists, and other visitors who want a special experience. Often when people meet me I can sense that they wonder, 'Who is this guy?' They want Francis or at least an *anglo*. By the end of the tour, however, they realize that I have given them facts, history, and made the tour relevant to what is going on in American wine, and they are impressed. I have so many letters that these people have sent thanking me. My daughter says I get these letters exactly because of their first impression."

Rafael talks about the importance of another vineyard on the property—the Gio Vineyard, named by the vineyard workers in honor of Francis and Eleanor Coppola's firstborn son, who died at age 23 in a boating accident.

"I always wanted to make the Gio Vineyard the best vineyard on the property because it was named by the workers for the Coppola's son, Giovanni. Gio was such a special young man; there was something amazing about him.

"Now I hope that someone will feel that way about Rafael's Vineyard, the vineyard named for me. I will try to plant something in that vineyard to make it special. There is some Niebaum clone of Cabernet on that land, and perhaps I can plant some of that in the Rafael Vineyard. The Coppolas really surprised me. They could not have honored me more. That was nice."

Rafael closes his eyes and says again, "That was nice."

And what of dreams not realized?

"I did want my own vineyard, but I think it's too late. I have had to put it out of my mind. I feel that this is my vineyard, and that certain wines are my wines. I feel that the 1954 and 1955 Inglenook Cabernet was my wine. The '82, '86, '90, and '91 Rubicon is close to my heart.

"If I had my life to do over again, I know that this is what I would do, but better. I would put into practice what I have learned, and that is that you have to treat people equally, love, and respect them."

Looking around the elaborate public space, the Inglenook château and its grounds, its fountains, its almost grotesquely beautiful staircase, its tasting room and gift shop, Rafael, who realizes that visitors to the château spend some money on wine or a t-shirt or a book, and spread the word about the Niebaum-Coppola wines, seems to take it all in stride, and not without some pride.

"The wineries that visitors to the Napa Valley visit—this is *numero uno*, then Mondavi, then Beringer. Opus One will be part of this tourist elite. Ninety percent of the visitors to the valley come here. It's amazing that so many people come, want a t-shirt, taste some wine. I think they want to be close to Francis, and they get so excited when they see him sitting quietly, drinking an espresso or a glass of wine. His name is very strong. Rubicon lives in the shadow of the Coppola name, and has found its place in the market."

With an open-to-the-public policy have come a lot of new employees at Niebaum-Coppola, most of them young, many of them ignorant about the history of Inglenook and what it means to American wine. Rafael does as much as he can to bring new employees into the fold.

"Not all of the new people working here feel attached to the property; for most of them it is just a job. Nobody working in the public area has been here even five years, and this is a problem. We have to watch some of these people, and see if they get that feeling, and then we have to encourage that feeling in them.

"I've been here a long time and can share my enthusiasm with these new people. Every month, I meet with new employees, and tell them the history of the place to light a fire under them. I like to think that there is someone among them that will take over the spirit of the place."

Reflecting on his own future, Rafael asks himself a rhetorical question. "When I retire, what will I do? I guess I'll come over here every day and visit Inglenook. I never had a hobby; my hobby was here. Tila is not a traveler; she is a real homebody, connected to the children and the grandchildren. Francis has said I can stay here as long as I want to, and he hopes that will be a good long time."

Because he must run to meet a VIP tour group of wine importers, Rafael Rodriguez has time to make but one more observation about Inglenook/Niebaum-Coppola, and the family that, after a thirty-year painful fissure, made the property whole. Looking at the limo that is delivering his tour guests, Rafael speaks softly.

"It's becoming a little showy here, but I know Francis and Eleanor will put the brakes on a bit. Niebaum-Coppola is attractive to so many people because of Francis. He is proud to be a member of the movie industry, and to have become an important member of the wine industry. It is amazing to see this man reaching for such heights of respect in the Napa wine community that others have taken lifetimes to achieve. He was very fortunate to become the owner of this property. A touch of luck, a touch of fate."

Rafael is standing at the main entrance of the Inglenook château, just outside the ornate Niebaum wine-tasting room. Even as he begins to greet the wine importers, who look surprised and bewildered when they meet the vineyard historian, Rafael Rodriguez asks a question that resonates within the massive stone walls of the Niebaum-Coppola museum.

"Just like Gustave Niebaum, Francis Coppola could go anywhere, anyplace. Why did he choose this place?"

No Winners

I have a thousand ugly stories."

Dennis Fife, the last president of Inglenook-Napa Valley, is talking about what he witnessed during his years as an employee of Heublein, the corporation that owned Inglenook from 1969 to 1990, but was best known for selling many millions of cases of Smirnoff Vodka every year. Fife, who began his career with Heublein in 1974, was general manager and president of Inglenook from 1983 to 1989.

Heublein, based in Hartford, Connecticut, wanted to become a major player in the California wine business. The company did so by acquiring Inglenook and BV (Beaulieu Vineyards), the two jewels of Rutherford. The acquisitions could not have been more different, however, in terms of Heublein's approach and intentions. In 1969, Heublein bought 82 percent of United Vintners, including Inglenook, for which it paid a total of almost $100 million. Later that year, Heublein bought Beaulieu Vineyards from the de Pins family for $8.5 million, just five years after United Vintners and its affiliated growers cooperative, Allied Grape Growers, purchased Inglenook from John Daniel for $1.2 million.

In one year, Heublein acquired the two most heralded wineries in the Napa Valley: Inglenook and BV. Heublein was a large, rich corporation, floating on a sea of Smirnoff Vodka profits. By all accounts, Heublein's chairman, Stuart Watson, was enamored of the image of fine wine and the Napa Valley fine wine culture, and argued in favor of purchasing both Inglenook/United Vintners and BV, even though Heublein's board of directors was dubious about the purchase.

Watson did not seem to be guided by lust for money or by the fear that Louis Petri expressed about the Gallo brothers buying the finest properties in the Napa Valley. Stuart Watson loved the image of Heublein owning these wine estates, and since the purchase of Inglenook/United Vintners and BV were small potatoes on the corporate ledger, and since Heublein was awash in Smirnoff-generated cash, the Heublein board indulged its chairman's whimsy.

Surely, the new owners would maintain the status of BV and mercifully return Inglenook to what it only recently had been: the finest estate wine producer in the history of California.

BV, yes. Inglenook, no. Why?

In no particular order: greed, ignorance, bad luck, bureaucracy, stupidity, lies, hiring, firing, distance, inexperience, corporate culture, indecision, mistrust, incompetence, jealousy, lawsuits, impatience, mentality. . . .

The list of buzzwords is endless, but Dennis Fife, who was at various times Heublein's vice president of their fine wines division, the head of sales and marketing for BV, and ultimately, the president of Inglenook, can set the stage to begin to explain how Heublein supported BV but squandered Inglenook.

"If I was writing the worst possible scenario for Inglenook, and this was clearly the worst, I never in my wildest dreams would ever have thought it possible. I mean, look, it's 1969, and you're Heublein, and you own both BV and Inglenook. You should own the fine wine business in the Napa Valley. What happened?"

Dennis Fife is fifty-four years old, and is the proprietor of Fife Vineyards in the Napa Valley and the Konrad Winery in Mendocino. He is known for making extraordinarily rich and complex Zinfandel and Petite Sirah wines in a world that cries out for more cheap Merlot and Chardonnay. Clearly, he loves running his own show. His severance package from Heublein allowed him to start his own wine business, and he is relieved not to be part of corporate wine production and marketing.

"Heublein was my first, last, and only corporate job. I'm not going back. I discussed the possibility of taking the job as president of Kendall-Jackson,* but I said no to the corporate world."

But Dennis Fife had not always said "no" to corporate America. As a matter of fact, when he was a younger man, he embraced the corporate world, and began to move up the corporate ladder quickly.

"I went to Berkeley and the Stanford Business School. A friend I grew up with in central California, Phil Baxter, became a winemaker in Napa Valley just at the time when the first American wine boom was starting. Phil got me hooked on enjoying good wine, and when I finished at

---

* Kendall-Jackson is a major player in the California wine business, and owns several brands, many of them formerly independent producers: Stonestreet, Edmeades, Cardinale, Robert Pepi, Kristone, and La Crema, among others. The company's flagship is Kendall-Jackson Proprietor's Reserve Chardonnay, with 8 million cases produced annually. The company, known for its aggressive corporate style, is owned by Jess Jackson, the noted San Francisco trial lawyer.

Stanford I knew I wanted to get involved in the wine business. I decided I didn't want to work for a big company, so I went to work for a small wine marketing company, and they went bankrupt three months later. I quickly revised my opinion of only working for small start-up companies.

"Heublein was looking for someone to work in what they called market analysis. Six months out of Stanford I went to work there. This was 1974, and I was thirty years old.

"At that time, Heublein was an independent company, later to be bought by larger companies; first, R.J. Reynolds (now RJR/Nabisco) acquired Heublein in 1982, and then they sold Heublein to Grand Metropolitan in 1987. I was a little fish in a very big pond.

"What I found interesting was that I was one of the very few people at Heublein who was passionate about the fine wine business. Most people had worked on potato chips or something like that in their previous jobs. So my interest and broad knowledge worked to my advantage. Every three or four months I got a promotion to another job—this was in San Francisco, the home office was in Connecticut—and I eventually became vice president for planning and development of the wine group."

As Dennis tells the story of his corporate ascendancy, he pauses for a moment and hits upon what he believes to be the basic reasons why BV and Inglenook were treated so differently by their corporate parent. For anyone who understands even a little bit about the dichotomy between the mammoth spirits industry and the simply large wine industry, his statement provides a crystallizing moment of clarity.

"Heublein acquired BV as a direct acquisition, but acquired Inglenook by acquiring United Vintners, which had bought Inglenook from John Daniel, Jr. about four years earlier. The federal government immediately slapped a Federal Trade Commission antitrust lawsuit on Heublein, so BV was removed from the wine division and put in the spirits division, operated by Smirnoff. BV was not put in the wine division until 1980, ten years later.

"Of course, all the money was in the spirits division, so BV got a lot of capital infusion, and not much corporate scrutiny. Inglenook was in the wine division, which represented a teeny piece of the company. I remember that the cigarette-lighter budget for the spirits division was larger than the total cost of the acquisition of BV, $8.5 million. Smirnoff Vodka was something like 22 million cases a year when I was there, with a cash flow of $100 million per year, and heavy profit margins: 22 million cases. That's more cases than the entire American wine business, excluding Gallo, so you're in a whole different world.

"How could Inglenook be any different from BV? You have the same land in the same place in the same valley, the same margins for product, so how come one made money and the other didn't, other than the way they were treated? BV got the capital infusion it needed when Smirnoff owned it. Inglenook had to get new equipment, but it was seen as a money pit."

From the beginning, Heublein's ownership of Inglenook was a dramatic mismatch that, was it not so sad and had it not ended so badly, might be considered comical. The people in Connecticut who made the important decisions had no idea of the value, prestige, and importance of Inglenook in the American wine business, and their decisions were misguided, inept, foolish, sometimes even vicious.

The bottom line, however, is that fine wine in general and Inglenook in particular never fit into the Heublein corporate culture, which was based on a commodity, spirits.

"It's easy to criticize Heublein for a lot of things, including the way they ran Inglenook, but the most legitimate criticism is that they should never have owned it in the first place. It was so different from their brands that it didn't make sense. Even today, Heublein still owns BV, but it just shouldn't be in the wine business at all."

The story of Dennis Fife's tenure at Inglenook is the story of someone who, because he was passionate about the history of the property and the lineage of its wines, tried desperately to change the culture of a corporation, and failed. He knew that Inglenook represented both the past and the future of wine in the Napa Valley, but he could not impress that fact on Heublein. His efforts are, to this day, fondly remembered by those who knew his commitment to Inglenook and the corporate difficulties he encountered.

Robin Lail, the daughter of John Daniel, Jr., and a person whose heart will always be connected to Inglenook, her childhood home, is one person who knew and appreciated Dennis Fife's passion for the heritage of the property and its wines.

"Dennis Fife did a first-rate job of attempting to bring that property back to what it was. He assembled a team of people during that time that were in the old Inglenook camp, including winemaker John Richburg.*

"Dennis took enormous corporate risks to do what he did. It was a first-rate effort, not just a good effort, and he accomplished many good things against almost impossible odds. So I have a warm spot in my heart for Dennis."

Dennis Fife was a passionate advocate for the heritage and quality of Inglenook Napa Valley, but he was working for a company that knew nothing about the business of making and selling fine wines. Worse, the company couldn't care less about any issues not directly tied to the short-term bottom line of Inglenook Napa Valley. He knew that as long as he tried to maintain the historic standards of quality at Inglenook, he would go nowhere in the Heublein corporation.

---

* Richburg, since 1993 the winemaker and a partner in Bayview Cellars, a Napa Valley boutique winery, worked for Inglenook Napa Valley for twenty years. He worked his first grape harvest when he was eight years old. Like Dennis Fife, John Richburg climbed the corporate ladder at Heublein, and when he retired he was director of winemaking for all of the company's prestige wines: Inglenook Napa Valley, Quail Ridge, BV, and the Gustave Niebaum Collection.

"I knew going in, because they actually told me that it was ok if I wanted to be at Inglenook, but if I was looking for a long-term career at Heublein, then this was not the place to be."

In retrospect, it seems odd that Fife, with an MBA from Stanford, might have failed to recognize the classic competitive strategic model that is taught to first-year MBA students. That model is: Neutralize a real or perceived threat (in this case, competition from other wineries), and then extract money from the asset (Inglenook) until it is rendered worthless. Then, move on to the next corporate conquest, the next site to slash and burn.

Was Dennis Fife so naïve or so lacking in business savvy that he missed the point of Heublein's strategic acquisition of Inglenook? It is more likely that Dennis Fife did recognize the textbook corporate strategy, but thought the asset of Inglenook to be outside the box of a classic model. Clearly, Fife believed that Inglenook needed to be preserved and nurtured, not destroyed. He must have also believed that he was the person who would open the eyes of Heublein to see the long-term value of Inglenook.

Fife is still haunted by what Inglenook might have been if only he did this or that differently. When you listen to Dennis Fife talk about Inglenook ten years after the fact, some of the wounds of those years seem fresh, some completely healed, but the impact of the experience on him is unmistakably palpable.

"I was the last president, the last corporate officer of Inglenook. For a long time—I felt guilty about it, the guy who presided over the end of Inglenook. Now my own business is so successful that I realize it's not me. What gets me is that Heublein did nothing, they ignored Inglenook completely. That was the story of my career there.

"It was like you could see the train coming, and there was nothing you could do. You'd say, 'Excuse me, there's a train coming, it'll be here in 45 minutes.' The reply would be 'There's no train.' And then, wham! You get run over."

Dennis Fife is talking about the man who taught him the fine wine business.

"If anybody was my mentor, Legh Knowles was. He was the general manager of BV and the most savvy person about the workings of the wine business that I ever met. My job, as director of sales and marketing at BV, was to translate what Legh said in fine wine language into corporate language, because the guys at Heublein didn't get it. They asked Legh once why he needed twenty fermenting tanks instead of ten, and he answered that 'we prefer to ferment our white wines separately from our red wines.'"

Legh Knowles, who died in 1997, was a legend in the Napa Valley, having run BV from 1962 until his retirement in 1989. A charming fellow, Knowles found his way into the wine business through a circuitous route; he had been a jazz and big band trumpet player. He learned how to sell and market wine at the hand of Ernest Gallo, for whom he worked for four years, and then was hired by the de Pins family to run BV. The Napa Valley and the French-owned Beaulieu were a far cry from Modesto and the hardscrabble world of the Gallos, but Legh Knowles made the transition gracefully.

Perhaps most important for BV is that Legh Knowles realized that BV estate should produce Cabernet Sauvignon almost exclusively, and he turned BV into the number one Cabernet producer in the Napa Valley. His reasoning was simple: BV had won so many awards for their Cabernet Sauvignon wines, especially the Georges de Latour Private Reserve made by André Tchelistcheff, and Cabernet was BV's most expensive and most sought-after product. Finally, and most important for Knowles, BV Cabernet Sauvignon was the wine that had the best story.

The best story? Dennis Fife explains why this meant so much to Legh Knowles.

"What I learned from Legh is this: you sell the wine and you sell the story. If you can't sell the story then you can't sell the wine. Lots of people make wine, and if you don't get the story through the sales force, you're doomed to failure. When I was at Inglenook, I remember that I went to Connecticut to meet one of my new bosses in sales. I knew that I was in trouble when his first question was 'What's your best-selling size?' He didn't even know he was in the fine wine business. He didn't have a clue. To succeed in this business you must have complete control over your sales force. No exceptions. It's like being a soldier: here's the job, now go and do it. Sell the story, sell the wine.

"There were distributors who had handled Inglenook for fifty or sixty years, and as soon as United Vintners took over Inglenook they were knocked out, and then Heublein knocked out their replacements. They fired all the most experienced salesmen because they weren't corporate enough or they weren't jug wine enough. They were really good guys, too. The sales force becomes all young, inexperienced kids that they can hire cheaply, and they don't know the story at all."

But surely Heublein knew how to sell wine? Their sales force sold all that vodka, all those spirits. They were a huge company, and even allowing for inefficiencies because of their size and Connecticut base of operations, the corporation must have understood the importance of a well-trained, cohesive, and motivated sales force. Dennis Fife explains that what he learned from Legh Knowles at BV fell on deaf ears at Heublein.

"I was at Inglenook for seven years. *In those seven years I had nine bosses and thirteen sales forces.* I never had a boss last more than a year, and only one sales force that lasted a year and a half, and that was the original one that Legh and I put together, and it was working. In one year, we went from 2,000 cases to 30,000 cases with the Smirnoff sales force. So the next year they told us to get our own sales force. It made no sense, except that Inglenook did not fit the corporate culture of Heublein, and was a corporate afterthought.

"But the corporate guys could party. They loved to come out to the Valley to relax and eat and drink wine. I remember that we had three and a half cases of the 1941 Inglenook Cabernet Sauvignon. In seven years, I might have used six bottles for the press. The corporate guys had an outdoor tasting and drank the rest."

In the December 31, 1990 issue of *The Wine Spectator*, a headline on page 16 reads: "'41 Inglenook Cabernet Sparks California Sale." The article details an auction at Butterfield & Butterfield in San Francisco, "where a bottle of 1941 Inglenook Cabernet Sauvignon, estimated to be worth $1,500, set a new single-bottle record for a California wine at the auction house when it sold for $1,800."

"Inglenook was treated like a stepchild."

Dennis Fife sounds more disappointed than angry as he recounts the evolution of the image of the Inglenook brand from a truly fine wine to a truly big jug.

"BV had the Smirnoff sales force and a blank check for their budget. Not Inglenook. The problem was not even the quality of our sales force. The problem was the sales force kept changing. Thirteen sales organizations in seven years. What that means is that you're irrelevant."

Was it that Heublein believed that Inglenook Napa Valley was a Tiffany-image money pit that was incapable of turning a profit?

"The fact is that Inglenook was larger and more profitable than BV until Heublein got their hands on it. If we had made more money than any other producer in the fine wine industry it still would not have made sense for Heublein to own Inglenook, because it was so intertwined with the jug wine corporate sensibility."

That jug wine was Inglenook Navalle, named for the Navalle Creek that Gustave Niebaum had resited to flow through the front of his property. Of course, the grapes for Navalle wines did not come from the Inglenook estate or even the Napa Valley, but from the high-yield–driven Central Valley. Navalle was one of many crosses that Dennis Fife had to bear during his tenure at Inglenook.

"I was president of Inglenook Napa Valley, so I wasn't involved with Navalle. I always felt that Inglenook Napa Valley needed to be separate from Navalle, and that Napa was dragged down by Navalle, because the public would not make the distinction between the two brands. We did research that showed it made no difference to consumers if the jug was sold as Léjon (another United Vintners brand acquired by Heublein) or as Inglenook Navalle, so naturally the wrong decision was made and for no good reason. One of the promises Heublein made to me when I took the job as president was that they would take the name 'Inglenook' off the Navalle label. They changed their minds and came out with an Inglenook Navalle Cabernet. What kind of sense does that make?

"Heublein didn't realize the value of Inglenook Napa Valley as a separate brand. Why? The culture. In the spirits business there are something like 100 brands that are worth $10 million or more, and they're coveted. Companies are always bidding on them. For example, Christian Brothers was acquired by Heublein for their commodity product, brandy.

"The idea in the spirits business is that all similar products make about the same profit per bottle, so it's simply a matter of volume to make more money. Heublein never understood that the fine wine business didn't work that way. There are wines that make $10 per case and wines that make $1,000 per case. They could not envision that they would reach their goal of Inglenook Napa Valley making a million cases of wine and make money."

Assuming for the moment that the number crunchers at Heublein might have been correct that Inglenook's fine wines would never turn a profit, was it a wise business decision to abandon the heritage of Inglenook and go head-to-head with Ernest and Julio Gallo? Could it be that Navalle could bail out Inglenook Napa Valley, allowing Fife to continue making fine wines? Dennis Fife smiles, then chuckles a bit, and then he answers these questions.

"The joke is Navalle made six million cases and no profits."

"Inglenook is a unique, great piece of land. There is so much diversity in one huge vineyard, so many different microclimates. The front vineyards where Niebaum moved the Navalle Creek, the back vineyards going toward the mountain, the vineyard surrounding the Coppola house — the soil is different in different places. I think if you make wine from the three properties that you make better wine than if you make it from any one of them. There is something magical about those three properties.

"I love all things that are handmade and have history, and here's Inglenook, a perfect example of that. We tried to bring some of that back to understand all the things that were great about the company. If you love the fine wine business, who wouldn't want to run Inglenook?"

One of Dennis Fife's achievements at Inglenook was the creation of "Reunion," a well-crafted Cabernet Sauvignon–based wine that was made from grapes grown on the original Inglenook vineyard (owned by Heublein and sold to Francis Coppola in 1995), the "home" vineyards (purchased by Coppola in 1975), and Napanook (a vineyard in Yountville that John Daniel bought in 1947 as a source of grapes for Inglenook. This is the vineyard that would later source Dominus.) The wine, made by John Richburg, was well received, as were many of the high-end estate wines from the Inglenook vineyards produced under Fife's supervision, and sold during that first year by the crack Smirnoff/BV sales force.

"After I was at Inglenook for a year we began to grow substantially in premium and ultra-premium wines: Chardonnay, Cabernet, Reunion, and items that hadn't grown in fifteen years, like Charbono.

"I remember the 1982 Cask Cabernet was an excellent wine, and we had put 500 cases aside to sell later at about twice the original price. We were selling it for more than $300 per case, on an allocation basis, as fast as it got to the market. For some reason, management discounted the wine to $40 per case wholesale. Why? We were selling it at the higher price — eight times as much — as fast as anything, and yet they turn around and give it away. Why? I don't think they understood what they had.

"Thanks to Heublein, our budget was $400,000, and we were expected to make a profit of $100,000 per year, a 20 percent return, which was unreasonable. The amazing thing is that they had lost $400,000 the year before, and we lost only $25,000 after our first year. I was amazed we came that close to breaking even."

"What really frosted us at the time was we could have made a profit. We had 100,000 gallons of Chardonnay all set to bottle, but Heublein decided that they needed it for another program, so they took the juice away from us. *They changed the plan and then told us that we lost money.* That was constant."

The attitude of Heublein toward Inglenook was pretty easy to sum up for Dennis Fife.

"I worked for them for seventeen years, and I was dedicated to Inglenook and to BV, but that was not the way the company worked. They didn't care about grapes, they didn't care about fine wine; they cared about marketing the wine, and they didn't want to put any money into the brand. My job was not to confuse them with the facts. My job was to fix the problem, to make a million cases of wine and make ten million dollars."

The attitude of Heublein toward the history and heritage of Inglenook and the Napa Valley was equally dismissive, as best symbolized by the barrel-storage building they constructed on the Inglenook estate. Francis Coppola, who now owns it, has muttered openly about blowing the building up, *Apocalypse Now*–style. Dennis Fife agrees wholeheartedly that the building is an eyesore, but says that the Napa Valley got away easy.

"People criticize the storage building. It's an eyesore, and it blocks the view of the Inglenook chateau from Highway 29. What most people don't know was that was just one of seven buildings planned by Heublein. People who did know thought they were nuts. Maybe not nuts, but stupid. Heublein built the storage building there to save $100,000, but it cost them $300,000 to upgrade the windows, etc. — another false economy. The reason why there are big doors so close to the roof on that building was to link it to the six other proposed buildings by a catwalk. The strict parts of the Napa Planning Commission in force today were invented because of Heublein."

Things didn't get any better for Fife and for Inglenook when Heublein was sold to R.J. Reynolds (now RJR/Nabisco) in 1982. Of course, Inglenook didn't figure much in Heublein's decision to sell, according to Fife.

"The reason Heublein sold was that they were having so much trouble with Kentucky Fried Chicken (now KFC). They didn't have the money to fend off a hostile takeover, so they went looking for a friendly buyout."

When RJR purchased Heublein, the corporate culture was based on the sales of hundreds of millions of cigarettes, not fine wine, or *any* wine.

"RJR culture was tobacco first and foremost. The guy running the wine division, just one of my nine bosses in seven years, came over from tobacco, and knew nothing about wine. He was in charge of BV and Inglenook. How could that be the right move?

"If the world were different, RJR could have been great for our wine business because of their advanced R & D. They bought a rose company because it was the first plant or flower company to use gene-splicing, and they wanted to gene-splice tobacco. Talk about clonal selection! But we were at the wrong end of the company. At the same time I was begging for a $1 million total budget, they were building a $1 trillion cigarette plant."

If RJR didn't understand the Napa Valley wine business, they certainly didn't understand the burgeoning Napa Valley real estate boom either.

"They did stuff like sell vineyards at $4,000 per acre, which in the three years they owned it had grown to a value of $16,000 per acre, so they sold it for $4,000 an acre. It was ignorance, not some grand plan."

R.J. Reynolds recognized the clash of cultures and in 1987 sold Heublein, including Inglenook and BV, to Grand Metropolitan, a slash-and-burn British megaconglomerate. And how did Dennis Fife, president of Inglenook, hear of the sale of the company?

"The night before Inglenook was sold to Grand Met the number two guy at RJR gave a speech about RJR's commitment to the wine division. When we woke up the next morning, the newspaper headlines told us we were sold."

Now, Inglenook became immediately expendable. Grand Met had Burger King, Pillsbury, Haagen-Dazs, and interests in the brewing giants Guinness and Whitbread. Inglenook was just red ink on their balance sheet; a salable brand name.

This was the beginning of the end for Inglenook, which was breathing its last gasps as a Napa Valley wine producer. While wine had not been made in the Inglenook château since 1964, when United Vintners disassembled the winery and moved it to Oakville, at least Dennis Fife was able to continue to make the wine in the Napa Valley from Napa Valley grapes.

Grand Met decided to stop investing in Inglenook Napa Valley, but even this final corporate nail in the Inglenook coffin was hammered in secret.

"I left just before they stopped making wine at Inglenook Napa Valley. Heublein, which was now owned by the Brits at Grand Met, said they were going to 'pull the plug' on Inglenook. They kept their plan to stop making Inglenook in the Napa Valley a secret from me, the president, for over a year. How dumb can you be?

"Up to a week before they started ripping apart the company, they were telling me they were going to double the size of our operation. If you can't make your mind up any better than that, then I don't know what. Maybe they were hoping for a miracle or maybe they had really made their minds up to tear the company down and I was the last to know."

Fife left Inglenook Napa Valley in October 1989 at the age of forty-three, and Grand Met began in earnest to put an end to the dreams of Gustave Niebaum, John Daniel, Jr., and Inglenook Napa Valley's last president, Dennis Fife.

In 1992, Inglenook Napa Valley's production facilities were sold to Sutter Home Winery, famous as the leading producers of White Zinfandel, who own them to this day.

In 1994, Grand Met sold the Inglenook brand, but not the estate, to Canandaigua Wine Company, based in New York, the second largest producer of wine in America, eclipsed only by Gallo. To the current generation of wine drinkers, Inglenook is just another brand of inexpenslve generic and varietal wines, most often sold for well under $10 in 1.5 liter magnums.

Dennis Fife points out the final irony in the history of Inglenook, once Napa Valley's jewel, now just a jug wine brand name.

"You know, Inglenook Napa Valley was never really sold. Navalle was sold and this little Napa Valley brand came with it. This is really amazing, because Heublein/Grand Met had turned down offers of $50 million for Inglenook Napa Valley. Essentially, they lost $50 million in the sale, all because of the way they went about it. Ridiculous."

Although he knows that he tried to make Inglenook the great Napa Valley estate wine brand it had once been, and he is pleased to be making his own high-quality wines in Napa and Mendocino, free of corporate interference, Dennis Fife can't seem to shake the imprint of that complex corporate experience.

"I still have dreams at night about convincing the head of Heublein that we're not doing it right; we have to do this and that differently. If I'm honest with myself I realize I wasn't at a high enough level to even ask for, say, three months to prove something to them, but I'm still frustrated. Why couldn't I change his mind?

"Even when I left, Dick Maher (then president of Heublein Fine Wine Group, who resigned in 1992) said, 'Dennis, we're not going to have presidents at this level, but there's always interest in you in Connecticut (at Heublein's corporate headquarters).

"I said that the only job I was interested in was president of Inglenook Napa Valley."

Dennis Fife wishes that he could have saved Inglenook, but he knows that is no longer an option, if, in multinational corporate America, it ever was. That fact does not diminish his enthusiasm for what is still possible at the original Inglenook Napa Valley estate, now Niebaum-Coppola. He is a keen observer of what Francis Coppola is trying to achieve, and he has strong feelings about the path that Coppola has chosen.

"I'd say 75 percent of what Francis has done I agree with, and 25 percent, most of it in the marketing, I would not have done that. I think the Rosso and Bianco table wines are the wrong direction with the wrong product. I think Inglenook needs a fanatic who wants to make the best wine in the world. If Francis says that's him, why would I believe him when he's taken bulk wines, overpriced them at $12, and they're just ok? The association of these wines with Rubicon hurts the new Inglenook. I wouldn't have done it.

"The more Francis can tie his wines to the history of Inglenook, like the Cask Cabernet, as a consumer and as someone who was tied up with Inglenook, I'll appreciate.

"To me, Inglenook was a Picasso that somebody defamed and destroyed, and if someone can now restore it, bring back those resources and that wonderful history, that person, to me, is a saint. And I think Francis will do that. He did it right as far as the property is concerned; the grapes first, then the winery. Francis has done a lot for the community, too."

Dennis Fife, who has been talking nonstop about his love for a very special place, his very personal angst over its corporate destruction, and his passionate hope for its renewal, falls silent for a moment, and then makes a final terse observation about the future of the estate he will always refer to as "Inglenook."

"Francis Coppola is the only guy who can put Inglenook back together now. He and Eleanor understand what they have. Heublein never understood Inglenook."

6

The Echo of the Vine

$\mathcal{T}$he French call it *terroir*. An elusive, indefinable mix of the soil, the mesoclimates and microclimates, the rainfall, the wind currents, all of the forces of nature. These ecological imperatives harmoniously conspire with human beings who respect those natural forces, and together produce a wine that speaks to a sense of place.

The French, the Italians, the Spanish have no word for "winemaker"; the closest near-synonym is the French *vigneron*, which translates as "wine grower," essentially someone who respects the character and heritage of the land and wants to make wine that sings its praises.

Officially, Scott McLeod is the winemaker at Niebaum-Coppola, a position he has held since 1991, when he was twenty-nine years old. More and more Scott finds himself roaming the 110 acres of vineyards, tending the vines, tasting the fruit. He knows that the best wines are made in the vineyard, not the cellar, and that the grapes grown on the Coppola Estate, grown on the vines of Inglenook, have an historic pedigree that sets them apart. He is searching for that pedigree, that character, those flavors that will allow the property to produce a wine that speaks volumes about the place it comes from, and the people who nurture and celebrate that place.

It is about 2 P.M. on August 5th, and the Napa Valley is enduring a week of uncommonly hot weather. It is 101°F in the vineyards of Niebaum-Coppola, and Scott McLeod is walking through the seemingly endless rows of Cabernet Sauvignon grapes, block by block, vineyard by vineyard. He stops every minute or so to sample a bunch of grapes. He knows that the 1997 harvest is about six weeks away, and so he spends several hours a day tasting, tasting, tasting. He is searching for flavors. He is seeking balance.

Scott tastes a grape at the bottom of the cluster, then a grape at the top, then a grape shaded from the sun by the leaf canopy, then a grape that is exposed, sunburned. He chews the bitter skins to check the tannin quality, spitting out the pulp and pits. Then, he spits the skin out, takes another grape, and concentrates on the sweet flavors of the juice. Tasting grapes this way, you soon begin to recognize the nexus of flavors and textures that can only be loosely defined as "complexity."

Often, the people who must work within the confines of nature wish that they could control it, even though they know and agree that nature will have its way in the *terroir*-driven vineyard. Scott, who, like so many of his fellow Napa winemakers is a graduate of the fermentation science/wine-making program at the University of California at Davis, is concerned about the effect of the torturous heat on the vines and the flavors of the grapes.

"If we didn't have this heat, the vine might work more efficiently. Photosynthesis loves 75 to 87 degrees. On a day like this—100 degrees—a dry-farmed vine will shut down and stop working at about 11 A.M. The vine requires moisture to operate and the stomata will shut the vine down as it operates in this moisture-losing heat. Just touch the leaves and you'll feel that they're warm. That means photosynthesis has stopped; the vine is not working."

Ripeness, which is based on sugar concentration, is almost never a problem for these grapes because of the power of the Napa sun, but maintaining fruit acids in the grapes requires reasonably cool weather. Without a balance of sugar and acid, the wine will be flat, because our palates will not be refreshed. This makes for a dull wine that will only get duller with age. In California, winemakers cannot chaptalize—add sugar to the juice to raise alcohol levels—but acidulation, the addition of acid to grape juice or to finished wine, is legal everywhere in the world. Even so, Scott much prefers to let these grapes speak for themselves in the finished wine.

"In California, too much sun is more of a threat than too much rain. Too much sun can ripen grapes too quickly, reduce color accumulation in the skins and, of course, lower acidity. I spend a lot more time worrying about sunburn than I do about rain damage at harvest. Loss of water concentrates both sugar and acid, but such high heat as we have today, as we've had all this week, destroys acid. Even though the grape is still acidic, the vine burns acidity, and so the pH goes up."

Water has a pH of 7, which means it has a balance of hydronium ions and hydroxide ions. pH in grapes is mainly an acid, water, sugar mix. In wine, the pH controls bacterial stability, color, and a bit of flavor. Scott likes to see a pH of about 3.5 in the Cabernet Sauvignon. The concentration of sugars should measure about 23° on a scale called "Brix," which will ferment out to about 13% alcohol in a dry wine.

"You can buy acidity, but you can't buy flavor. So, we still try to keep our pH for Cab at 3.5 in this kind of heat. pH rises, so we have to pick on flavor. In the back of our minds, we worry about acid to a point, but we can buy acid. We can't buy flavor. The flavors are everything."

How does Scott taste for flavor? What does he look for? What flavors are the unique expressions that define the Inglenook/Niebaum-Coppola vineyards?

"I always look for negatives first—green pepper, bell pepper—and sometimes that never burns off if the temperature is not high enough or the wind is too much. A little bit of bell pepper makes it interesting, but too much makes it taste green, even though the fruit is ripe. A desirable vegetal characteristic is just a hint of green pepper, not a lot of it, and not a lot of black olive or oregano spice. We made one wine on this property in '95 that tasted like black olives. In the fermenter it smelled just like a pepperoni pizza. We ended up not using it in the Rubicon because the wine didn't taste like the property.

"What does Cab taste like when it's ready to pick? A depth to the fruit, kind of briary, and because of the acid level, it tastes like jam, rather than fresh fruit. The flavor is reduced down, a bit warmed but not cooked. Chew the skin and you taste the bitter tannins come out, and you can tell when the tannins are ripe; they're a bit soft.

"I like it when the red fruit character just begins to show in the skin, and the color of the seeds change quite a bit, very brown when the grape is perfectly ripe. I know when the taste is right that the seed will be brown.

"Cab is too easy, because it colors so well. Zinfandel, you have to track the color throughout the berry; wild strawberry when completely ripe, blueberry when it's not quite ready."

As he walks slowly through the Niebaum-Coppola vineyards, tasting and touching and looking, and focusing all of his senses on the soil, the vines, the fruit, Scott McLeod is a winegrower. He is not *making* wine, because he, like everyone at Niebaum-Coppola, is not interested in producing "just another California Cabernet Sauvignon." He is *growing* wine, because in the history and evolution of this estate, it is Scott McLeod's moment to give voice to the unique and unquantifiable character of its vineyards, creating wines that sing arias, hymns, scat, the gospel, and wines that rock on about the history, the heritage, the place that is Inglenook/Niebaum-Coppola.

Some people, including academics, wine producers, and grape growers, dismiss the idea of *terroir* as it applies to California. There are those who believe that because the vineyards of California are not as old as, say, Burgundy (some of Burgundy's vineyards predate Charlemagne, who ordered the monks to grow grapes for wine in order to support the monasteries of the Holy Roman Empire), and so the vineyards cannot lay claim to a site-specific *terroir*. Others believe that the whole idea of *terroir* is romantic nonsense, and that differences in the flavors of California wines can be explained by science, and that the character that is most important in a wine is varietal character—the character of a particular grape, not the particular site.

Scott McLeod, who believes in and understands the *terroir* of Inglenook/Niebaum-Coppola, articulates that *terroir* with such elegance that it is hard not to become a believer, hard not to see what he sees, hard not to taste what he tastes.

"The older the vine the less it reacts to seasonal variations. The older the vines get on this property the closer the flavors get within the same spectrum, within the same family."

"You taste fifteen-year-old vines, not the original Niebaum clone, but they begin to approach the property flavor, the property specificity—the *terroir* of Inglenook. You can taste it in the finished wines."

"When the wines from older and younger vines, from different vineyards, different blocks, are tasted together before blending, the wines structurally seem to have a lot of similarity: the sweetness of the core; the way the fruit finished and continued to finish for several minutes; or the sweetness or ripeness of tannins. Flavors are different, but there is something intrinsically similar about the wines, and the only thing they have in common is the property, a great growing site.

"There's got to be a *terroir* to this property. The property expresses itself as the wine ages. There's a cedar-spice element in the nose of the wine that develops as the wine gets older. When I taste young Cab I get violet or almost a floral, youthful character. Austere tannins and a wild blueberry/violet core, that is not a typical Cab.

"As the wines age, they approach a oneness in the bouquet, in the aroma. It is the cedar-spicy aroma that is really defined through the older vintages. It comes up after about seven years and I don't know where it comes from. Children don't really look like their parents until they get older.

"Overall, you've got to believe that if you grow these vines on one site long enough, there's got to be something that defines the wine. I think it's easy to believe that you have *terroir* when you don't. Anyone who grows anything wants that, whether it's tomatoes or grapes."

Scott becomes animated when he deconstructs the idea that it is the clone of a vine—a plant propagated by the use of cuttings or grafting from the original plant to maintain its character and distinctive traits—that accounts for a wine's character. Inglenook is best known for the Niebaum clone of Cabernet Sauvignon, cuttings that Niebaum brought from Bordeaux to Rutherford, and have been propagated on the property since the 1880s. The Niebaum clone has been "cleaned up" by the plant pathologists at the University of California at Davis, meaning that it has undergone a flash-heat treatment and certified as essentially virus-free. The Niebaum clone of Cabernet Sauvignon is the backbone of Rubicon, and to listen to Scott McLeod, this historic vine provides the genetic material to make a Grand Wine.

"What separates Rubicon from our other wines is the clonal factor of Cab. The personality of the Niebaum clone comes through in the wine. It smells like violets, as opposed to raspberries or cassis, and makes a wine that can be pretty tough and austere, but more often gorgeous. It always needs to be tempered with other Cab from the property, but it's got that distinctive Niebaum signature, a stamp.

"All the Cab clones in California are great; they all express something different. Our Niebaum clone, however, makes our wines taste unique, and it is grown only on this property."

Niebaum is not the only clone planted on the property, however. The property has plenty of Cabernet Sauvignon Clone 7, from Concannon Vineyards in the Livermore Valley, and certified by UC Davis. Clone 7 is known for producing high yields in the vineyard and wines with true "varietal character"; the wine made from Clone 7 is Cab and tastes like Cab.

The University of California at Davis has been instrumental in the growth, modernization, and success of the California wine industry; UC Davis is famous worldwide for training aspiring winemakers by teaching them the science and technology of viticulture and viniculture. Some say that this education comes at the cost of winemaking traditions, while others feel that knowing the science is a way to explain and more fully grasp the reasons for those traditions. Whatever the perception of the Davis education, the fact is that the university turns out a talented and enthusiastic crop of winemakers employed mostly in the high-tech California wine industry.

In the late 1950s, the United States Department of Agriculture and the California Department of Food and Agriculture began Grapevine Certification and Registration, an ambitious program to provide grape growers with vine stocks free of viruses, such as leafroll and fanleaf, and resistant to fungal diseases, such as eutypa and powdery mildew (also known as oidium). The UC Davis department of plant pathology developed systems—

"indexing" and "heat therapy"—for selecting vines to meet these criteria, and the university's Foundation for Seed and Plant Material Service distributed a "mother block" of virus-free grape varietals to registered nurseries, who sold the approved stock to grape growers throughout California.

During the 1960s, Dr. Harold Olmo, a plant geneticist at UC Davis, was instrumental in the development of laboratory clones of the Cabernet Sauvignon vine. The purpose of these Davis-certified clones was to increase yield and to prevent disease in the plants, which would lead to healthier and more productive vineyards. Olmo and his associates mostly worked on Cabernet clones that would do well in the hot and humid San Joaquin Valley, but several of these same clones, especially Clone 7, have been planted throughout California, including the Napa Valley. Currently, Clone 4, with plant material coming from Argentina, is enjoying some popularity at Niebaum-Coppola's neighbor, BV. Clone 4 is a vigorous producer and makes a complex wine.

While most viticulturists and wine producers agree that the Davis certification program has had a positive impact overall, a significant minority of quality-driven producers argue that the disease-free and fungus-resistant vines are far too vigorous. The vines produce a lot of fruit, but not excellent fruit with balanced flavors. These same critics note that these viral and fungal diseases add to the stress on each vine, and this is a positive aspect of the life cycle of the vineyard.

Scott McLeod, Davis graduate, believes that the work the university has done has been important, but ironically leads him back to his argument for *terroir*. McLeod observes that producing a certified, standardized vine has not led to standardized aromas and flavors in the grapes produced from those Cabernet clones.

"Look at the same clone of the same grapes, say Clone 7 of Cabernet Sauvignon. Well, it's the most widely planted clone of Cab in California.

When you try it in Calistoga, or here, or Monterey, or wherever, you realize that the clone of the grape is far less important than the place it's grown.

"*Terroir is everything*, and people who don't believe it are foolish or haven't spent enough time proving that wine from different places tastes different, property to property, and that's an indisputable fact. There's plenty of Clone 7 planted throughout California. Why doesn't it all taste alike?"

"This is an historic Cabernet property, and since Francis and Eleanor want to restore its greatness of its previous years, it's always in the back of our minds that this property is about Cabernet. I think it's crazy to throw more than 100 vintages of great Cab experience away, and I think it's just good business to grow Cab in western Rutherford."

Scott McLeod is talking about his passion for the wines produced from this property, and as usual, he tempers his reverence for the property's history with an eye to the current and future wine market realities that he deals with on a daily basis. Scott's conversational view of history and commerce seems an unconscious attempt to bring some grounded and contemporary balance to the idealistic and historic Grand Design of Francis Coppola: to produce one of the world's "Grand Wines"—a fine wine with the heritage, history, and *terroir* of this property. After all, it is Scott who has to make the Grand Design work in the vineyards and cellars of Niebaum-Coppola.

So, Scott is a practical guy, but that in no way diminishes the respect he has for the dreams of the Coppolas or the dreamers of the Old Napa Valley. He admires all of the dreamers, and when he talks about them, he is protective.

"One huundred years ago you had Inglenook, BV, Beringer, Krug, Greystone, Far Niente, Trefethen. It's incredible to think that in this very narrow window in the history of this valley that those entities were created. Captain Niebaum was a legend among legends. His dream was not just a personal dream 100 years ago, but knowingly or unknowingly, a financial dream. He made Inglenook a classic institution. So, I take heart when Francis says that he 'consults the Captain' in making decisions for the property.

"The fact that Inglenook had been so plundered in the past by Heublein, it just makes Francis and Eleanor's commitment to the property all the greater. Heublein looked at Inglenook as a brand, not a property, and they knew that Inglenook was known for Cab, so they decided 'Screw the property, let's buy some Cab, and produce more than the 20,000 cases the property can turn out.' That was the first crime, and after you commit the first crime, the second and third crimes become easier to commit.

"The second crime was to stop making estate wine on the property. The third crime was to say, 'The hell with the property'; sell it and keep Inglenook as just a brand, with nothing but its name to back it up."

Scott McLeod has seen the effects of slash-and-burn corporate viticulture in these vineyards, and with Rafael Rodriguez and Rafael's crew, has brought the vineyards back to life through replanting new vines, and replanting an old mindset: grow the grapes for the best possible flavor, not the highest possible yield.

Scott knows what he wants from this soil and from these vines and he knows what Francis Coppola wants from his wines, especially Rubicon. Scott and Francis have a goal for Rubicon, which with the opulent 1994 and 1995 vintages, they may be close to attaining on a sustained basis. It is Francis' idea of a Grand Wine, as articulated by Scott, in cooperation with the *terroir* of the estate.

Coppola defines a Grand Wine as "a wine that can please contemporary taste, but it has to have an historical aspect, and the vineyards must be at the zenith of what defines an area."

Making reference to the "First Growths," the *Premier Grand Cru Classé* wines of Bordeaux (Margaux, Lafite-Rothschild, Latour, Haut-Brion, Mouton-Rothschild), but applying that concept to California, Scott McLeod identifies the group of wines that Rubicon must match and surpass if the Niebaum-Coppola estate is to realize its historic potential and the destiny of its heritage.

"The number one thing is that we want to compete with the First Growths of California, all Cabernet-based. They are Opus One, Mondavi Reserve Cabernet, Heitz Martha's Vineyard, Phelps Insignia, BV Private Reserve, etc. Some are First Growths because they are classic wines, some due to marketing and public acceptance, and the field is expanding." (Author's note: I would also include Shafer Hillside Select, Diamond Creek, Mayacamas, Louis Martini Monte Rosso (from Sonoma), and Château Montelena in this group.)

"This property can create a First Growth in Rubicon; previous incarnations were excellent, although some are quite rustic. In the early days of Rubicon, when André Tchelistcheff consulted on the wine, we used to hold back the wine seven years, so we looked at it differently. Francis wanted the wines to be ready, but stylistically the wines have changed."

Varietal character is the mere baseline for any wine bearing the name of a grape on its label. Cabernet Sauvignon from Chile, Australia, or California that sells for $8 has varietal character, and this is all the consumer should expect from a wine at that price point. A "First Growth" Cabernet, however, must exhibit far more than that. It must speak to the vintage year: the soil conditions, the weather patterns, the time of harvest, and the choices made in the vineyards and winery during that vintage.

The wine should also announce its address, the place that it comes from, its *terroir*.

"Some of our wines you do have to wait for, and what you wait for is a vintage characteristic, a property characteristic, not a varietal characteristic. We are looking for lots of grip and lots of tannin, which is what the property provides. I would rather have someone *choose* to hold the wine five more years rather than feeling that he or she *has* to hold the wine five more years. Many of the wines from the 1980s were wines that you *had* to wait for."

André Tchelistcheff is revered throughout California and especially the Napa Valley for his work at BV, where he made Georges de Latour Private Reserve Cabernet Sauvignon, the Opus One of its day. After he left BV, he consulted for many wineries, including Niebaum-Coppola. If you are going to knock his style of Cabernet, you had better be able to say why, and back it up with something better. Scott McLeod is doing just that, and he is not alone.

Dennis Fife, who was president of Inglenook during part of the Heublein debacle, worked closely with Tchelistcheff, an inveterate smoker for all of his adult life — he lived to the age of ninety-two — because BV, Inglenook's closest neighbor, was also owned by the spirits behemoth. Fife amplifies McLeod's criticism of the old style of what was considered Napa's best Cabernet Sauvignon when he says, "Almost nobody at BV or Inglenook smoked, except André Tchelistcheff. He was a great taster; he would taste and smoke, and it was fine for him, but nobody else in the room with him could smell or taste the wine. I think his wines are the wines of a smoker's palate."

Fife makes an astute observation. The old BV Cabs and the old Rubicons have the bouquet and flavor of once fine but now stale cigars, and the only way that any fresh, mature, or even dried fruit is going to emerge from those bottles is if, for the sake of palatal contrast, you burn your palate with cigarette, cigar, or pipe smoking. Like so many older wines, they are mostly disappointing curiosities.

What Scott McLeod is doing, what Tim Mondavi is doing, what Craig Williams, who makes Insignia at Joseph Phelps is doing, what Elias Fernandez and Doug Shafer are doing at Shafer, what Mike Martini is doing, what Bruce Cakebread is doing, what so many of the baby-boomer and post-boomer winemakers are doing in the Napa Valley seems anathema to the classic Tchelistcheff Cabernet model. That model was based on the Haut-Médoc district of Bordeaux, where they have great historical vineyards, great winegrowers and great wines, but not much sun. The wines of the Médoc emphasize youthful tannins, but the best always have a distinct background vein of mature fruit that emerges with age.

The old Napa model went too far, eschewing the ripeness of the sun-kissed fruit, in order to be taken seriously in the France vs. California/ Bordeaux vs. Napa boxing match. The wines were ponderous, emotionally dark, and we were told to lay them down, and those bitter tannins would somehow turn sweet, the vegetal bouquet and flavors would somehow turn to fruit, and the wine would last forever. The truth is that among some very fine wines (especially wines from the 1970s made by Robert Mondavi; the '73, '74, and '75 are stunning), there were many wines produced that, were they emperors, would be changing into their new clothes with every vintage.

So what are these boomer winemakers doing differently? Why are their wines so radically different from so many of their progenitors? What they are doing is celebrating the sunshine of the Napa Valley, making great wines full of ripe flavors that are approachable when young, and the wines will last. Winemakers like McLeod and those of his generation don't suffer from the Bordeaux Complex, from the French Fantasy. They are making sunny California wines with both ripe fruits and ripe tannins, picking grapes based on the flavors that their vineyards provide. Luscious, sexy American wines to appeal to people who just love great wines with great food. What is happening in the vineyards is what is making these wines exciting (in the winery they sometimes tend to overdo the new oak to please the vanilla-soaked palates of some American wine writers).

Since 1995 the retail prices of some of Napa Valley's finest wines have doubled or even tripled, but this has not diminished consumer demand. These wines are on allocation because they have proven so popular. Try to find a bottle of (among many others) 1994 or 1995 Rubicon, Opus One, Insignia, Mondavi Reserve, Diamond Creek, Caymus Special Selection, or Shafer Hillside Select on a retail shelf at any price. You will find them only on wine lists of wine-destination restaurants at heart-stopping prices. A 1994 Rubicon can easily bring $150 to $250 in any restaurant lucky enough to have a few bottles in its cellar.

How does Scott McLeod make this new style of Rubicon so that it is successful in the marketplace *and* fits Francis Coppola's definition of a Grand Wine? Clearly, Scott has thought this process through, and is happy to share his approach.

"We take a very good look at the relative ripeness of tannin and fruit before we pick based on flavor in the vineyard. When the wine is ten months old we put a trial Rubicon blend together, and think, 'What can we do, maybe 2 percent more cab franc?' At 20 to 22 months, 'What can we do to lift aroma?' Here, we can play with 5 percent of vintage plus varietal." (Legally, the vintage stated on the label must be 95 percent of what's in the bottle.)

"The other magical component of aging is not acid, tannin, or alcohol, but fruit extract, the total fruit impact on the wine. You can make great, delicious, fabulously silky, fruit-driven wine, where the fruit is not so narrowly defined as a bushel of strawberries, but rather the wine has this underlying core of concentrated flavor, as opposed to structure.

"Rubicon is built for the long term, but in a balanced way. If the wine is not balanced at the beginning of its journey, why would it be in balance twenty years later? What do we expect to happen along the path that's going to push it one way or the other?

"All tannins are not equal. Very ripe tannins diminish with age, but nasty, harsh, bitter tannins do not diminish, and that's the worst thing; you have a fall in the ripe tannins, and a fall in the luscious fruit flavors as the wine ages, but goddamn if you still don't have those nasty tannins in the wine.

"So, if you have a short, hot year, like '96, I want to get less tannin out of the skin. Last year, a short crop and hot weather, you've got all this leaf power ripening less fruit, and because of the heat it's maturing faster. The goal is to make sure you have less overall tannin at thirty days then you did at twenty days or fourteen days. The fact that the wine is good is exciting."

Among adult Americans who drink alcohol (about 50 percent of that population), consumption is only 2.5 gallons of wine per capita, and that consumption level is not rising. Compare this to the French and Italians, who consume 18 gallons per capita, but this includes the entire population of each country: every man, woman, and child. Americans are actually drinking less but better wines, as sales of premium wines have skyrocketed at the expense of the formerly reliable jug wine business.

Scott McLeod reveres the history and winemaking traditions of the Napa Valley, but he is also an active and enthusiastic player in the contemporary wine business. He knows that to sell premium American wines to American consumers you not only have to grow and make good wine, you have to be able to tell the story of that wine.

"In California, we sell the fact that wine is a romantic lifestyle. The truth is that it's hard, sweaty work, the competition is endless, and foreign imports increase every year. The wine business is business in its rawest form; we're talking about an agricultural product that is easy to ship and conserve, so the world is our competition. Shipping peppers or lettuce is a lot harder. What makes wine so romantic is that Americans are so separated from the winemaking process, and it's not a daily product as it is in Europe. When you sell it as a specialty product, there has to be an element of romance to it."

So, are the processes of creating the wine and creating the market for the wine mutually contradictory? Is one process "hard, sweaty work" and the other process just good storytelling? Not at Niebaum-Coppola, since Scott doesn't totally separate one process from the other.

"There is no contradiction between making and marketing our wines, and I'm involved in our marketing. We don't want Rubicon, for example, marketed in a way that is outside of its context as an estate-produced premium wine. I think the whole system works best when the wine is marketed in the proper context, and that is, in the case of Rubicon, that it is made from the historic Niebaum clone of Cabernet Sauvignon, grown solely on this estate, and it is the best wine this property can produce.

"I think we can improve our wine as much as 5 to 10 percent, but we're a year or so from having ideal temperature and humidity in our cellars. The winery has started to make money, and that money is going into the cellars and into building a new and modern winery."

Gustave Niebaum knew that he had to produce the finest wine in this country under the best conditions in the vineyard and the winery in order to sell a product he was proud of to the public. Both Coppola and McLeod have taken a page from Niebaum's book, rejecting the modern notion of marketing first, quality second—if at all. The quality of the wines drives the marketing, and the success of the marketing of the wine dictates higher and higher quality in the wines.

"Francis has risked a lot to save Inglenook, and the wine is great, but you have to be as meticulous in the cellar as you are in the vineyard to make the wine we can and will make. So, marketing the wine, selling the wine, will help to make the wine better and better."

Certainly, Francis Coppola has become increasingly and heavily involved in the marketing of Rubicon and the other wines produced by Niebaum-Coppola. Francis, because he is a famous filmmaker-turned-wine producer, provides the media with the sizzle, and Rubicon is the steak (though Francis insists that "Rubicon is both the sizzle and the steak"). He sits for interviews, meets the press at the Inglenook château, attends wine industry dinners where Rubicon is served, and has become a ubiquitous presence in food- and wine-related print and electronic media.

Scott McLeod is happy to see Francis becoming more involved in marketing the wines, because it helps to complete the circle which begins with an idea. And that idea comes from Francis Coppola, not from a marketing executive or agency.

"Francis is an idea person, so he loves new ideas. He gives us ideas and we have to decipher how to turn that idea into a new wine, for example. We have to make the wine, but we also have to figure out how best to sell that wine. Like right now, Francis is focused on producing some Aglianico (a red grape native to Basilicata, reflecting Coppola's southern Italian heritage). So I'll plant an acre and we'll make the wine from it; perhaps just for Francis, perhaps not. It may become another estate wine for sales and marketing."

"Francis is focused on producing some Aglianico" barely begins to describe the seemingly obsessive nature of Coppola's new idea. For three months I spent part of every day with Scott McLeod and we would run into Francis on a regular basis. Every time he spoke with Scott, often before saying hello, another part of the Aglianico idea would emerge. It became clear that in Coppola's mind it was a very short jump from *talking* about the idea of planting Aglianico and producing the Napa Valley's finest estate-grown Aglianico wine (a category that Coppola would own;

there are no other producers of varietal Aglianico in Napa), to actually carrying out the idea.

Once the Aglianico idea was complete, it appeared that in Francis Coppola's mind the wine was already made, waiting to be enjoyed. Of course, the Coppola Aglianico will have a great story, full of Godfather-like Italian and Italian-American imagery. I would imagine that it might be the next wine in the Edizione Pennino line, named for Coppola's maternal grandfather's music publishing company, which is now represented by estate-grown Zinfandel. Zinfandel is, in fact, a clone of the Primitivo grape native to Italy.

If the wine is any good at all, and there is no reason to think that it won't be very good, Pennino Aglianico, when coupled with the story, should be as hard to obtain as the very fine Pennino Zinfandel, which is sold on allocation.

The idea that came several ideas before Aglianico was Cask Cabernet, an idea that came to fruition. In late 1998, 700 cases of 1995 Niebaum-Coppola Cask Cabernet were released. Inglenook, in the days of John Daniel, made its reputation on bottlings of cask-designated (e.g., Cask J-12) estate-grown Cabernet Sauvignon, and Francis, looking to the history and heritage of the property and winery, wanted to make an homage to Inglenook and John Daniel. So, he came up with the idea of Cask Cabernet.

Francis, while not involved in the growing of the grapes or the making of the wine, made everybody crazy over the "Cask" label. He wanted the greatest label possible to represent this wine, and the concept of the wine: estate Cabernet Sauvignon whose initial release celebrates the year in which he purchased the Inglenook château—1995—and celebrates the history and tradition of the finest cask-designated wines of Inglenook. It is, in all likelihood, the most expensive wine label ever produced (the original Rubicon labels were, at Coppola's insistence, printed by Tiffany's, who thought they were printing bar mitzvah invitations). Of course, the "Cask" label is dramatically beautiful. Of course, the label is made of wood.

Scott McLeod seems to have embraced Coppola's creative method, as it allows him to make a Grand Wine—Rubicon, a *Meritage* blend of Niebaum Clone Cabernet Sauvignon, Merlot, and Cabernet Franc—and now, a Great American Wine—Cask Cabernet—100 percent estate Cabernet Sauvignon.

"Francis came up with the idea for Cask, and we discussed it for about a year and a half before we went ahead with the project. I've always been interested in making a wine that expresses California, but not a *Meritage* style. I have always wanted to make a 100 percent Cab. It won't have the property stamp that Rubicon has, since it will use far less of the Niebaum clone than Rubicon. The wine will go into cask, a high percentage of new American oak. I want to make a gutsy, delicious, flavorful Cab. We are starting with '95 vintage, and by the '97 vintage, released in the year 2000, I expect to have enough to really start producing for the market."

Looking across to a bucolic hillside vineyard that will source much of the 1995 "Cask," Scott McLeod begins to tell the story of the wine, starting in the vineyard.

"This is the old Inglenook cask lot, a gorgeous vineyard, a beautiful setting. It was an eight-acre section of Cab planted in 1936 where Inglenook got all of its Cask Cabernet and so we'll call the vineyard "Cask." Rafael (Rodriguez) has always called it Pritchard Hill. When I started working here eight years ago, this block was bare; it was taken out in 1991, and we replanted in the spring of '93. We let it recover and we redeveloped the entire plot on a new rootstock, 110R.

"I see my primary job as getting this property to the point where it grows the Napa Valley's greatest Cabernet Sauvignon; that's its history and its future."

Isn't this a potential marketing nightmare: Rubicon vs. Cask Cabernet? Scott is aware of what he calls "marketing challenges," but seems excited by the prospect of creating his own competition.

"Quite a few of the wineries in the valley have made their reputation with consumers by good varietal wines with identifiable characteristics. Rubicon is a wine that you have to search a bit, investigate a little. If you were to serve Rubicon and Cask to consumers, probably 50 percent would prefer Cask, and 50 percent Rubicon. Since Rubicon is made from the best stuff on earth, the Niebaum clone, and based on the fact that there's only 4,000 cases, you don't want it to be too popular.

"Cask is going to be so good that even people who don't like Cab are going to like Cask. It's labeled as a Cab, which will be a first for us. It's fruit forward, luscious, ripe, sweet Cab fruit, since younger vines, due to their vigor and shallow roots, can't produce complexity and longevity, but they sure can produce sweetness.

"Rubicon is identified closely with this particular property, and it is a wine that we make for the world stage. We will make sure that Cask is about making Rutherford Cabernet Sauvignon—a wine to celebrate Rutherford—which is the appellation on the label, same as Rubicon. Cakebread, Sequoia Grove, BV, Turnbull, Grgich Hills, Whitehall, Caymus, are all Rutherford, and we want to promote Rutherford. We're very bullish on Rutherford as the greatest place in the Napa Valley for growing Cab. I want people to ask questions when they taste Rubicon, but Cask is going to be just a great Cab."

Some of the questions that McLeod probably wants people to ask about Rubicon might take the form of an interior dialogue, and include thoughts about the winemaker's intentions, about the property from which it comes, about its longevity, about its quality. He very likely wants people to question if it is among the finest wines they have ever tasted, and to affirm that indeed it is.

As a young winemaker, Scott McLeod wants to continue to challenge himself to make the finest wine possible from the cherished historic vineyards of Inglenook/Niebaum-Coppola. For him, that wine is unquestionably Rubicon. He realizes that making great Cask Cabernet, Merlot, and Zinfandel from this property and successfully marketing all of these wines will not only raise the standard for Rubicon, but for all of the wines he makes. Success will make his job harder, and that is the challenge he seeks.

"Rubicon always has to be a scarce product. If we have 500 tons of Cab from the estate, and 150 go to Rubicon, then 70 percent will go to Cask and only 25 to 30 percent to Rubicon. If Cask nips at Rubicon, and Pennino nips at Cask, and Merlot nips at Pennino, then the competition among wines from this property will raise the bar for all of our wines."

The idea that came several ideas before Cask, which came several ideas before Aglianico, was the idea for "Francis Coppola Presents" wines. Produced as *Rosso* and *Bianco,* they are wines that sell for $10, have colorful labels of the vineyard-as-stage-set, and are sourced from purchased grapes grown in California, mostly in the Central Valley. These are the most controversial wines identified with Coppola, as the very concept of making jug wines—and five to ten years ago these wines would have been sold in jugs—is anathema to what Francis and Eleanor Coppola claim they are doing: giving new life to the most esteemed estate in the Napa Valley. Rosso and Bianco are not estate wines, they are not Rutherford wines, they are not Napa Valley wines, they are not North Coast wines, they are "California" wines (actually, the Rosso tastes a lot like an Italian wine from Emilia-Romagna visiting California for the first time).

The 1997 Bianco is a palatable blend of Chardonnay, Muscat Canelli, and Riesling grapes, while the 1996 Rosso, a good wine with pizza or pasta, is made from Zinfandel, Syrah, Petite Sirah, and Carignan. By calling them Bianco and Rosso, Scott McLeod does not get locked into blending the same grapes in the same percentages vintage after vintage. He can make the wines to fit a set of established taste parameters within the price point he has to meet.

The initial bottling of '95 Rosso and '96 Bianco—15,000 cases—sold out quickly, and now McLeod purchases enough grapes and juice to make about 35,000 cases per year. The wine is made and bottled at rented facilities in St. Helena, because the tanks and equipment at Niebaum-Coppola's winery can just barely accommodate the estate wine.

Rosso and Bianco and the whole "Francis Coppola Presents" concept is controversial because it begs the question, "How are these wines any different from the inexpensive varietal jugs that Heublein turned out and now Canandaigua turns out under the Inglenook label?"

The differences between Coppola and Heublein or Coppola and Canandaigua are many, but can be summed up pretty easily. The previous corporate owners saw the Inglenook estate as an expensive liability to get off its balance sheet. They probably believe that wine labeled "Inglenook" can be produced anywhere—it doesn't even have to be California; it can be Chile or Argentina—just put the stuff in a bottle and sell the brand name.

Perhaps the answer to the question lies in the fact that Coppola has a vision for Niebaum-Coppola that is a sustainable vision, not the slash-and-burn vision of the previous corporate owners. He wants very much for the estate to thrive, based on the quality of the wines produced from its vineyards. Yes, he is producing jug wines whose sales are nudged by the fame of his name. At the same time, Coppola is also intent on producing what he has termed a Grand Wine, and that is Rubicon, followed closely by Cask Cabernet. Francis Coppola Presents a Means to an End.

The usual scenario for producing a cash-cow wine from an upscale winery is to make what is known as a "fighting varietal," a drinkable, but hardly remarkable Chardonnay, Merlot, or other varietal-labeled wine that sells for less than $10. And make a lot of it. Fetzer makes hundreds of thousands of cases of such wines, like its Sundial Chardonnay. Robert Mondavi produces millions of cases of the very successful Woodbridge line of fighting varietals. Gallo makes Gossamer Bay, Turning Leaf, Zabaco, and some Gallo-Sonoma varietals at this price point.

Scott McLeod and Francis Coppola wanted to take a different approach to making an affordable wine for everyday drinking that would improve cash flow for the company. For Francis, it was a celebration of the simple, inexpensive, often homemade wines that his family enjoyed. For Scott, it was to make enjoyable, fun wine, but without a real identity tied to grape or place.

"I see making Rosso and Bianco as far more exciting than making just another mediocre California Chardonnay or Cabernet. Our wines are fun, they're successful, and they're not varietals. We have no more estate grapes, so we created Rosso and Bianco.

"I'd rather make a Rosso that doesn't relate to anything about this property, or a particular varietal grown here, or anything that has to do with the quality of the property. I don't think there's a strong identity with the property with Bianco and Rosso, because it's 'Francis Coppola Presents,' not 'Niebaum-Coppola' on the label, just as there isn't that strong an identification with Rubicon and Francis."

Scott also makes the estate-grown Francis Coppola Family wines, which have gotten good reviews in the wine press. The Chardonnay, Cabernet Franc, and Merlot appear on wine lists in many upscale restaurants, and do well in retail outlets. The 1994 Merlot received a rating of

94 out of 100 points from the *Wine Spectator* and was considered dramatically underpriced, a steal at $24. One thousand cases, the entire Merlot production, sold out in six weeks. The 1995 vintage sells at retail for about $34. The wine is delicious.

The Bianco/Rosso program has created a buffer for McLeod, so that the winemaker does not have to sacrifice the quality of the Family wines to make more money. Making jug wine also allows McLeod to subvert the context of the modern California wine industry a little.

"A financial consultant would say, 'Let's see, you made 1,000 cases of Merlot and sold it in six weeks. Why don't you make more Merlot?' Economically, that makes sense. However, we'd rather make more Rosso and Bianco as a workhorse wine than to flog a mediocre Merlot. If you get more wine from the same amount of grapes, something's got to give; the wine quality will suffer.

"We're kind of stupid, I guess. We've got a lot of Cabernet Franc that tastes like a decent Merlot, and is possibly a clone of Merlot; it's easy to misidentify them. Nursery Cab Franc looks quite different from Merlot, but visually you can't tell the difference in the field clone of Franc from Merlot. We could make a lot more money selling it as Merlot, but since it tastes like truly great Cab Franc, we're truer to the consumer, truer to the property, and that's important to us."

Scott McLeod has written his indelible signature on winemaking at Niebaum-Coppola. His estate wines capture what is best about the place the wine comes from, the character of the vintage, and the flavor profile of the grape varieties planted on this historic ground. He is passionate about growing and making wine at Niebaum-Coppola, but what about his own future?

"If I'm going to make Cabernet that is true to its site in the Napa Valley, I can't imagine a better place on earth to be. Francis rewards those who stay with him. Our future, the future of Niebaum-Coppola, is bright, so that means my personal future is bright. My time and my focus have changed from winemaking to the vineyard. The fact that we grow and make it all—I know what they're going to be like as we grow and make them, vine by vine. We have confidence in the way we make our wines.

"How did Francis come into the Valley and take away the most significant Cabernet site in California from industry veterans and professionals? I really don't know. It's certainly more than money. Francis and Eleanor never had any intention of owning this property; the property spoke to them, and now they are committed to it 100 percent. And so am I."

Harvest Time

*E*rle Martin is vice president and general manager of Niebaum-Coppola, and as such is responsible for making the Coppola wine business work on a day-to-day basis. Like winemaker Scott McLeod, Martin is in his mid-30s, but unlike most everybody else at the winery, he is first and foremost a businessman, an entrepreneur. Erle does, however, have a thoroughgoing, almost envious grasp of the mechanics of the fine wine business, and because of that fact, he knows that he is in charge of the business end of a winery that has the potential to produce the finest wine in the United States.

Erle Martin broke into the wine industry in 1986, straight out of Boston University graduate school, and chose to undertake a trial by fire. He went to work for Gallo, training as a sales rep, and went through Gallo's management training program for three months, at what is often called the "University of Modesto." Gallo is famous for its take-no-prisoners sales and marketing force.

Ernest Gallo, the ninety-year-old patriarch of the company, has always been in charge of sales and marketing at the family-owned company. He was instrumental in designing a training program that has become the stuff of legends in corporate America. Gallo-trained sales reps are known to come into a liquor store or supermarket and rearrange all the wines on the shelves, so that Gallo products are at eye level. They have also been known to hide or even remove the products of competitors from store shelves.

In her 1993 book, *Blood and Wine: The Unauthorized Story of the Gallo Wine Empire,* author Ellen Hawkes reveals some early Gallo marketing tactics that strain credulity, including Gallo sales reps smashing bottles of their cheap, fortified "street wines," like Thunderbird, Gypsy Rose, and Night Train in the gutters of the nation's poverty-stricken neighborhoods, so that men sleeping it off on city streets would see the shattered products upon waking. Talk about brand imprinting.

Hawkes also reveals that Ernest and Julio Gallo, and their younger brother, Joseph, Jr., were orphaned in 1933. Their father murdered their mother with a shotgun and then turned the shotgun on himself. Ernest and Julio raised Joseph, who was thirteen when his parents died; he worked for the family company for nearly 30 years, until he established his own vineyards and dairy farms. He believed that Ernest and Julio started Gallo wines, that the older brothers were the owners, and he had no stake in the company.

In 1986, Joseph began to sell cheese under the Gallo name. He was promptly sued by his brothers for the illegal use of his own family name. It was during this court battle that Joseph learned that Ernest and Julio did not start the wine business by going to the library and borrowing books on winemaking, as the elder brothers have always maintained in their official company biographies. What became clear is that Joseph Gallo, Sr. left the remnants of a grape brokerage and winemaking business—complete with a secret underground wine production facility—and that a bootlegging uncle gave the young Gallos an assist in early distribution.

Joseph, Jr. countersued his brothers for his share of the wine business. He lost that suit, and he lost the right to put his own name on his
dairy products to Ernest and Julio, his brothers, and until Joseph was 18,
his legal guardians. If this is how he deals with his own brother, it is no
wonder Ernest Gallo has developed a reputation as a tough businessman.

Whatever one may think of the methods and ethics of Gallo's sales and
marketing program, the members of their sales force, still standing after
having gone through that trial by fire, are aggressively courted and
recruited by the entire California wine industry, including the boutique
wineries of the Napa Valley.

Erle Martin left Gallo in 1988 to work for Young's Market, a wine distribution company founded in 1888 (not long after Captain Niebaum purchased Inglenook) that had become in the ensuing 100 years, like Gallo, a
family-run billion dollar company.

"Young's presented me with a lot of opportunity, and I made the most
of it. I got there just as the company was reorienting itself, going from a
spirits-based house to a wine-based house. I rose out of my peer group to
become the wine sales manager for the state of California, and ended up
managing $75 million worth of business.

"It was fun for me to help shape the company and acquire definitive
brands—Iron Horse, Hess, Kenwood came on, and so did Fetzer, Kendall-
Jackson, Sebastiani, and quite a few small independent producers. I realized that what I was passionate about was brand building, which is not as
cut-and-dried as distribution. I love to take a brand, image it, grow it, and
make it powerful."

As part of his job at Young's Market, Erle pitched Niebaum-Coppola
to become part of Young's portfolio. He met with Francis Coppola and
came away impressed with the potential of Coppola's dream. He did not
get the business for Young's, but did get a handle on Coppola's approach
to the wine business.

"I knew that Francis was very different from the other wine producers I worked with, because after I made my unsuccessful pitch, I got a gracious thank you letter from him. Typical of Francis, in the body of the letter he started mining me for information. He's a great student; he'll go to anyone who he considers to be an expert in his or her field. He's deferential to people who know their business, and he'll try to pull out the information that he needs from them."

About eight months after their initial meeting, Erle Martin got an unexpected phone call from Francis Coppola's office in San Francisco.

"I was at the Four Seasons in Laguna Miguel, when I got the call. I was about to go home to Marina del Rey, so I left my mobile number. While I was driving, the phone rang, and it was Francis, so I pulled the car off the road. He caught me off guard, hit me between the eyes. He asked me point blank, 'What would it take for you to work for Niebaum-Coppola?'"

Erle Martin was hired in June 1996 as vice president of sales and marketing, essentially filling the role of national sales manager for Niebaum-Coppola. A year later, he was promoted to his current position, and although he seems rooted in the realities of the wine business, he is also aware that he is working for an unusual CEO.

"When I got here, the place was not in chaos, but there was a vacuum—a lot of people needed to know who was running the place, and so I did my best to learn all the facets of the organization. Now that I've settled in as the general manager, my hands are in everything, which in terms of my own personal growth is fantastic.

"It can be an overwhelming job, and I think I still have a lot to learn. My forte heretofore had been sales and sales management. I've expanded and I have the wholehearted support of Francis. I think I share in Francis' vision. We're constrained by the same things that other people are constrained by—money—but when you sign on to work for Francis, you know that you're going to be competing with his vision. My burden is to take his concepts and make them practical; work out the mechanics. By looking at Niebaum-Coppola as a business and evaluating it that way, I believe I'm taking Francis' vision and moving it forward."

How does Erle move that vision forward while still trying to create a healthy business that will sustain itself now and in the future? His answer, characteristically, refers to an established academic model that he can apply in the real world.

"In grad school, you learn that any product has a four-part life: introduction, growth, maturity, and decline. At some point, all products go through that. When I came here, Niebaum-Coppola was breaking that model in that it introduced, grew, matured, and was kind of topping out quickly. And of course, the entire thrust of the place was geared to make Rubicon.

"Now, however, things are changing as we are in a phase of reintroduction of the products of Niebaum-Coppola. We're at a very dynamic place in the history of the winery, because we have new wines, we have the Inglenook château, tens of thousands of visitors annually. We're much more public, and that's a big change."

Talking with Erle Martin, you are struck by his holistic view of the wine industry, and where he sees Niebaum-Coppola positioned in that industry. He prides himself on having the most current information on virtually every aspect of global wine commerce. Unlike Niebaum-Coppola's neighbor, BV, which is still owned by Heublein/Grand Met, and managed largely *in absentia*, Erle Martin is a very hands-on manager. He makes sure that he knows what is going on in every aspect of vineyard and winery operations, and is persuasive in molding those operations to fit what he considers to be sound business principles. The applications of these princi-ples are designed to serve the financial interests of the company he man-ages, serving those interests now and in the future.

Erle Martin seems comfortable in a business suit, and when he repre-sents the company in public, his image is impeccable, his corporate rap tai-lored to his audience. When he works at his office in the Chiles House, adjacent to the Inglenook château, he sports designer casual attire. He is well groomed, photogenic, and quick-witted—the very image of a future CEO. Erle sees his mission at Niebaum-Coppola as lighting a fire under everyone, including Francis, to make the business healthy, and to expand the brands of Coppola's wine business.

It would be easy to regard Erle Martin with the combination of respect and tolerance bordering on dismissal that many people accord aspiring corporate climbers, but it would be wrong to do so. Listen to Martin and you realize that you are listening to a dreamer. His dreams for the future of Niebaum-Coppola are not so different from those of Francis and Eleanor Coppola, or Scott McLeod, or even Rafael Rodriguez. Certainly, his pas-sion for Niebaum-Coppola is unequaled, even if he is able to express that passion in terms that the corporate world can understand and support.

"Three years ago the general public thought Francis was a Hollywood guy making wine. The insiders would say, 'God, there's a lot of potential there. Kind of a sleeping giant, a little quirky, marketing's out of sync.' They wanted to know: 'What's the plan? Who's manning the ship? Who do you want to be?' There wasn't any strong definition. Francis' name was not being leveraged for the good of the winery."

Erle Martin, by quoting Niebaum-Coppola's critics, has rapidly defined the elements of the strategy that he has put in place to answer those criticisms, first with opening gambits, then with definitive salvos that resound throughout the Napa Valley and beyond.

"Today, it's quite different. People are all over us. They've seen the publicity, they see the château, they taste the recent and forthcoming wines, and they're blown away. People want to be with us. They see us as the upcoming star of the Napa Valley. Both quality and marketing are up to snuff, and catching attention."

Becoming more specific, Martin talks about how recent activities at Niebaum-Coppola have helped to focus the public and wine industry consciousness on the company. He uses the unsettling but somehow appropriate metaphor of firing a blast from a gun to wake people up.

"Shot 1: acquisition of the new property, reuniting the original estate in 1995 and the release of Scott McLeod's first Rubicon, the 1991, is released in March of 1996. Shot 2: the '92 Rubicon gets good press, Parker's on the bandwagon,* and now the Merlot gets attention, and I'm sure the Pennino Zin is going to be fanned a little brighter. We make the best estate Zinfandel in Napa. Shot 3: the opening of the Inglenook château to the public, and the public loves it. Sales of wine in the château every month surpass sales from the previous month. We do from $250,000 to $400,000 in sales per month during summer and fall, and wine sales are about 60 percent of the mix. Shot 4: the release and success of Rosso and Bianco."

For all his fired-up enthusiasm for public acceptance and recognition of Niebaum-Coppola as a brand, and the embrace of the wine industry and press, Erle Martin never forgets Francis Coppola's vision, his dream, to make Rubicon a Grand Wine.

---

* Robert Parker writes the most influential wine newsletter in the United States, and his ratings, based exclusively on his own palate, can make or break a fine wine. Parker did a vertical tasting of Rubicon with Coppola, Martin, and McLeod on the verandah of the Coppola home, and was impressed.

"A big part of my desire is to see this winery earn its rightful place in the minds of wine connoisseurs of the world; I don't really care about collectors. There's a tremendous amount of pride, a kind of pride of ownership among the people who work here; an unbridled energy and enthusiasm. I think that this winery should make one of the finest wines in America. Everything we need to make that wine is here, or will be here. Sure, we need a new winery, and we'll get one. We have opportunities to do things that other people can't.

"We want to produce the best wines, and to be recognized for what we do, not because it's Francis Ford Coppola the celebrity, but Francis Ford Coppola the passionate artist. The wine and winemaking are extensions of his passion and his artistic influence on the property and on all of us."

Erle Martin appears to be the ideal choice for general manager of Niebaum-Coppola. He buys into Francis Coppola's vision for Niebaum-Coppola, but based on the realities of a dynamic world wine market. Martin, who would likely be the first person to define himself as an aggressive business person and creative and insightful marketer, is not immune to dreaming dreams as grand as Francis Coppola's. Erle would probably call his dreams "strategic planning."

Erle Martin respects and admires winemaker Scott McLeod for taking Rubicon and the other estate wines "to the next level." He feels that it is important to recognize Scott "for the great winemaker that he is; he's the guy that's making it happen on a day-to-day basis." When Erle Martin talks about Rubicon it is clear that he has an educated palate, and has tasted thousands of wines from all over the world.

"Scott has developed a style of Rubicon that expresses the property; he's not trying to bring Bordeaux to Rutherford. He's making a new style of wine that celebrates the fruit of these vineyards. Rubicon '94 is a mind-blowing bottle of wine. Maybe one out of twenty people like the old style of Rubicon as compared to Scott's. It's not as refined and rich as Opus One or Caymus, and not as rough-hewn as Diamond Creek or Maya-camas. It's a wine that is true to the property, a marketable wine that doesn't pander to the wine critics.

"People get caught up in tasting wine as a cocktail beverage, or they feel they have to score it, taste oak, taste tannin, anything but enjoy it as a beverage with food. I've tasted some of the big oak bombs. They're wines for the uninitiated. People want to be hit over the head with these flavors, and so do many wine writers who can't evaluate wine that isn't loaded up with oak. Scott knows that anyone can buy new French oak barrels, but he's looking at each grape as a constituent of the wine."

Erle Martin genuinely believes that Rubicon is a world-class wine. Still, Erle knows that it is his job to position Rubicon as a Grand Wine not only in the eyes of Coppola and McLeod, but also on the palates of the public and wine writers. As the flagship wine of the company, Rubicon must be a success for the other pieces of the Niebaum-Coppola wine puzzle to fall into place. This is a challenging balancing act that can make the difference between Rubicon becoming an overwhelming success in the fine wine market, or labeling Rubicon forever as an interesting, honorable, and expensive failure.

"The downside of the wine press is that everybody waits for the experts to tell them what to drink. People need to rely on their own taste. I never want to infringe on Scott's vision for Rubicon, but I have to balance his passion with practicality. I want to make sure that we don't develop a 'house palate,' so we'll do competitive tastings, see what other people are doing, and I bring Scott feedback from the trade as well.

"Scott's a purist, and part of my job is to reign him in a bit, so that what he feels is right will, at the end of the year, be able to balance the books.

But the bottom line is that Scott's instincts are right. Rubicon, because of its source material, has the capacity to be one of the finest wines produced in the world, and thanks to Scott, we're right on the cusp of that."

By any measure, Rubicon has become an extraordinary wine. Tasting the '94 and barrel samples of the '95 confirms that it is one of California's greatest Cabernet Sauvignon wines (the '94 was 90 percent Cabernet Sauvignon, 5 percent Merlot, and 5 percent Cabernet Franc). The fruit sings in the wine, the elements of flavor, aroma, and bouquet achieve a delicate balance, and the wine's complexity assures longevity.

The 1995 Rubicon sells for about $80 at retail; the 1994 sold for $65; the 1993 vintage sold for $50. All three vintages are sold out, and so will most likely appreciate in value. The wine is not inexpensive, but is a good value in the truest sense of that word. The '95 Opus One, one of the Napa Valley's "First Growths," sells for $125. Both wines are truly extraordinary examples of the 1995 vintage in Napa, but only one of the wines is estate-grown and bottled, and only one of the wines speaks to a sense of place, heritage, and history: Rubicon.

So how can Erle Martin help to allow Rubicon to realize its potential? When you are making a fine wine, at a certain point, you have to recommit to the wine, which means putting money into the least sexy, almost invisible aspects of the winemaking process: improving and expanding the cellar, making barrel storage more accessible, better temperature and humidity control, new refrigerated tanks and hoses.

According to winemaker Scott McLeod, "We could make Rubicon better, ratchet it up maybe 5 to 10 percent if we build a new winery." Erle Martin knows this to be true, and knows he must come up with the money and the space for that winery, but that's not all. The winery has to not only meet Scott McLeod's standards for a great wine, it must reflect Francis Coppola's vision of the heritage and history of the Inglenook château as the place to make a Grand Wine. Francis wants Erle Martin to make it happen, and soon. Erle is all too aware of the impatience that sometimes accompanies Coppola's vision.

"We have to apply for a permit for a new winery, but Francis feels that the winery should be grandfathered, and the last thing you ever want to say to Francis is that it can't be done. He has lived his life, his success and his failures, by pursuing his heart. Francis thinks it's bullshit to have to get a permit, because this is Inglenook. The county board of supervisors doesn't see it that way.

"By hook or by crook we will be producing wine here in an ultra-modern state-of-the art facility, even if it is housed in a Victorian stone château. We have a higher storage than production capacity, and we need to change that. The new winery is the last part of the vision for Rubicon, and when it's done, it will be the first wine since 1964 produced in the Inglenook château. It will happen, and it will be amazing."

So, Erle Martin is bright enough to know that Rubicon is the golden goose of the company, and he is committed to allocating the resources from within and the interest from without to making it the Grand Wine that Coppola envisions and that McLeod knows he can grow and produce. Erle also knows that for the company to grow and prosper attention must be paid to making all of the assets of the company work synergistically, one building on the other. In Erle Martin's world of sales and marketing, Niebaum-Coppola's least tangible asset is the most valuable.

"One of the assets that we can leverage is Francis Coppola's name. It does legwork for us that few other wineries can do. Spinning off the "Francis Coppola Presents" Rosso and Bianco wines, there is a lot of opportunity for us to develop a line of "Francis Coppola Presents" products, such as sauces, foods, etc."

Also, Martin must consider the impact of the grapes grown in the vineyards purchased by Coppola in 1995 as part of the Inglenook acquisition, and the impact that fruit will have on estate wine production.

"This year, we'll ship about 40,000 cases total; about 15,000 of that will be estate wine, so we don't need to break down the organization into estate and non-estate wine yet. When all of our vineyards are pumping, we'll incrementally be able to produce between 55,000 and 60,000 cases of estate wine. At 60,000 cases, the estate will be self-supporting. We might want to create separate companies for non-estate wines, like Rosso and Bianco, and related products, and keep Niebaum-Coppola products tied only to the estate."

Speaking of the estate wines, Martin is aware that he must reposition the estate-grown Merlot, Cabernet Franc, Chardonnay, and Edizione Pennino Zinfandel in the market.

"Our Francis Coppola Family wines and Pennino do extraordinarily well in the wine press, and we notice that we've undershot our prices in the past. Our '94 Merlot got a 94 out of 100 points in the *Spectator*, and it sold for $24; we were the least expensive of all the Merlots with scores of 90 or above. That wine sold out in a matter of weeks. I could have priced it at $36 and no one would have batted an eye. We only make 1,200 cases of Merlot, 1,000 cases of Cabernet Franc, and not much more estate Zinfandel. Francis has always wanted to keep prices moderate, so we are going to have to raise prices modestly but stop undercutting ourselves."

The 1995 Francis Coppola Family Merlot sold for $36 and sold out as quickly as the '94.

Another of the marketing possibilities being explored by Erle Martin is the possibility of a "white Rubicon," a blend of Marsanne and Roussane—grapes native to the Rhône Valley—blended with Chardonnay. The white Rhône grapes, along with Viognier, another "Rhône Ranger," that makes up 4% of the Coppola Family Chardonnay, are all grown on the estate, and show some promise as the backbone for an ultrapremium estate wine. The one thing that may hold back this idea is that there is no historical precedent for such a wine from this property, and so the prospect of such a wine may have minimal appeal for Francis Coppola.

Erle Martin's immediate challenge, however, is the test-balloon marketing of Niebaum-Coppola's newest wine, which is rich in the property's history and heritage. In late 1998, 700 cases of 1995 Niebaum-Coppola Estate Cask Cabernet, complete with unique and expensive oak veneer labels, were released. The wine, which harkens back to the Inglenook Limited Cask Cabernet Sauvignon wines produced under the ownership of John Daniel, is 100% estate-grown Cabernet Sauvignon. In barrel tastings, the wine was very fine—lush red fruit, deep black cherry color, a delicious glass of wine—and because of clonal selection, vineyard location, and age of the vines, extraordinarily different from Rubicon. Cask Cabernet is sure to be compared to its progenitor. Erle Martin is, not surprisingly, enthusiastic about the new wine.

"We see the Cask Cabernet program building to a very substantial part of our business, eventually selling 15,000 to 20,000 cases of estate wine.

"Cask Cab will be released earlier, about two and a half years after vintage—Rubicon is released four years after vintage—and the wines will be stylistically different. Rubicon is the greatest achievement of this property, but Cask is also an important estate wine. So, Cask will not be subordinate to Rubicon by design. Neither wine is market-driven; they are property-driven. So Cask should raise the bar for Rubicon."

"Francis is more than a celebrity, he's an icon."

Erle Martin is talking about working for Francis Coppola, and like many of the employees of Niebaum-Coppola, Martin imbues Coppola with the larger-than-life image that Coppola, consciously or unconsciously, seems to cultivate.

The difference between Erle Martin and other winery employees is that Erle has just as grand a vision as Francis, but the focus of that vision is vastly different from Coppola's. For Erle Martin, doing an excellent job for Francis Coppola is highly rewarding but not an end in itself.

Clearly, Erle Martin has taken an important step in his career in the wine industry by redefining, repositioning, and expanding Niebaum-Coppola as an important player in the world of wine commerce. He is given a great deal of independence in decision-making, and he has earned that independence. He understands the current and future sales and marketing needs and imperatives of Niebaum-Coppola and works very hard to attain them.

Erle Martin is a loyal and valued employee at Niebaum-Coppola. He does not, however, fill the same niche as winemaker Scott McLeod, who sees his future tied to the long-term growth of the vineyards and winery. Scott seems truly connected to the property, and it would be hard to imagine him *not* making Rubicon. It would be easy to imagine Erle moving up elsewhere in the wine industry; he certainly has the talent, ability, and drive.

Like Francis Coppola, Erle Martin appears to be incapable of doing one thing at a time; he likes his job because he is involved in every aspect of the business. But once he has achieved what he wants to achieve at Niebaum-Coppola, once his activities have become routine, it would be hard to envision Martin not moving on to another, bigger challenge. His own words paint a picture of a confident young man in a hurry to achieve success at the highest level, a CEO waiting to be anointed. Yet, Erle Martin is a man without blinders. He realizes that along the way to realizing his own dreams, his own grand vision, there are countless challenges and lots of hard work.

This Victorian house, built for Gustave and Susan Niebaum, has been home to the Coppola family since 1975. The Coppolas have maintained every original structural and architectural detail of the house, except one: a window was set in the kitchen.

*Fog rolls in from the San Pablo and San Francisco bays at night,
reaching Rutherford in the early morning hours. The cool mist provides
enough moisture for the vines and allows fruit acids to increase in the grapes.*

*A Cabernet Sauvignon vineyard at Niebaum-Coppola
facing the Stags Leap district of the Napa Valley.*

Rafael Rodriguez

Robin Lail

Dennis Fife

Scott McLeod

Cabernet Sauvignon grapes ripening at Niebaum-Coppola. The gnarly vine wood indicates that this is an old vine. Indeed, these grapes are the Niebaum clone of Cabernet, dry-farmed, without irrigation.

Grapes picked on the Niebaum-Coppola estate are received at the winery. The small machine in the foreground is a crusher/destemmer. The larger cylindrical machine is a bladder press, which presses the grapes gently to extract the sweetest juice. The staircase leads to a digital sound recording studio.

"Francis Coppola Presents" Bianco

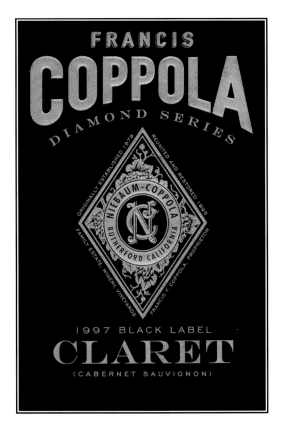

"Diamond Series" Black Label Claret

"Francis Coppola Family Wines" Merlot

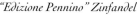

"Edizione Pennino" Zinfandel

Niebaum-Coppola Cask Cabernet

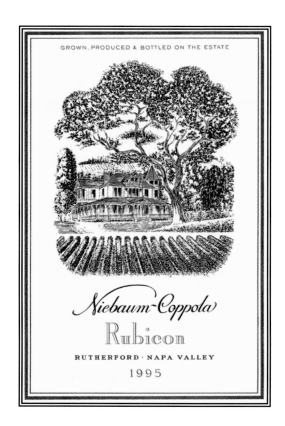

Niebaum-Coppola Rubicon

The Inglenook Château, built by Gustave Niebaum in 1887,
has been completely restored by Francis and Eleanor Coppola. Open to the public,
it now houses tasting rooms, shops, and a museum dedicated to two families:
Niebaum and Coppola.

*Francis and Eleanor Coppola enjoying lunch in the beautiful gardens of
the Niebaum-Coppola estate. The garden adjoins the family home.*

"Without question, this is a unique opportunity, and a lot of people would be envious of my position. But walk a mile in my shoes; 18-hour days in a dynamic environment, working for a dynamic individual.

"Francis is very good about empowering the people he hires, and I hope that Francis feels I am working hard and working smart and he is getting a good return on his investment in me. The flip side of all that is that people who work for Francis don't have real long careers, either. I'm not deluding myself. I don't think I'm going to be getting my gold watch here when I'm fifty."

8

Stewardship

*I*t's always fun to have lunch with someone who loves to eat," Eleanor Coppola observes, as her guest can't get enough of the exquisite meal served in the Coppola family gardens. She is clearly pleased that her attention to detail is noticed, and that the colors, aromas, and flavors of the food and wine all conspire with the beautiful surroundings to make for a relaxed and almost otherworldly repast.

The table is set with exquisite simplicity; flowers arranged in the center, surrounded by a Mediterranean feast. "These fresh anchovies preserved in olive oil are a gift from our neighbors, and so are these ripe peaches. I brought back these giant beans from Sicily — I forgot what you call them — and planted them here with great success. The tomatoes and lettuce are from the gardens, and the grapes are from our vines. I like to serve our Cabernet Franc with lunch; it tastes so good with this food."

The wine is a perfect match with this food, and the synergy of a wine made from one of the classic red grapes of Bordeaux paired with traditional peasant food of southern Italy is somehow emblematic of the Coppola's stewardship of the historic Inglenook vineyards and château.

The food is an extension of the Coppola family's roots, and the wine is an extension of the Coppola family's vision. In fact, the label reads "Francis Coppola Family Wines."

Now there are four labels: Niebaum-Coppola is reserved for Rubicon and the new Cask Cabernet; Francis Coppola Family Wines are varietal wines that are grown, produced, and bottled on the estate, including Chardonnay, Viognier, Merlot, Zinfandel, Cabernet Sauvignon, and the Cabernet Franc; Edizione Pennino, named for Francis Coppola's maternal grandfather and his music-publishing business, reserved for its extraordinary and hard-to-obtain single-vineyard Zinfandel; Francis Coppola Presents, a label created for moderately priced, California-Italian style *vino da tavola* wines, "Rosso" and "Bianco."

For Eleanor Coppola, overseeing and keeping track of all these lines of wines doesn't lead to marketing confusion, or a schizophrenic identity for the family's wine business. She sees the branching out of the different wines as many spurs emerging from the same vine.

"I think our dream is more grounded than it's ever been, or at least I can see the possibility of being grounded. The winery as a business is going forward into the next century, if we can keep the vision going forward."

When Eleanor Coppola speaks of "the vision," she credits Francis as the auteur of that vision. This quiet but confident modesty seems to be a product of living with someone who is perceived as being larger-than-life, someone who can articulate and execute big ideas, but who, at base, is someone who is creating something to honor his family and its posterity. She is comfortable with being consumed by both the grandiose and simple aspects of that vision, the way someone swept up in a cyclone believes everything will be all right as long as those she loves hang on.

"Francis works intuitively with an underlying sense of his own family values, though he doesn't do it consciously. Niebaum-Coppola, the land, the winery, the history, the business, is a greatly magnified expression of wine in his family. We know these grapes have great pedigree, and we set out to make the finest wine. You end up making a grand wine at the premium end of the wine spectrum, honoring the Inglenook history, honoring the families of Niebaum, Daniel, and Coppola."

The intuition of Francis Coppola is not so far removed from the instincts of Gustave Niebaum and John Daniel, neither of whom were professional winemakers. Inglenook at its best has always been the expression of the vision of passionate amateurs who loved and cherished the property, and in doing so honored the land and the heritage of that land. When Inglenook found its way into the hands of nearly invisible corporations, first with Allied Vintners and then with Heublein, the wine suffered, the land suffered, and certainly the heritage of Inglenook suffered. So, Francis Coppola's intuitive work is totally consistent with the best instincts of the families that he and his family wish to honor: Niebaum, Daniel, and Coppola.

The Coppolas are thrilled that Rubicon's reputation has grown exponentially, expressing the power of the place and people from which it comes. But like all people who really love wine and food, sometimes the Coppolas just want to enjoy a good bottle of wine that is not connected to the precious history and culture that is Inglenook.

"Francis always wanted to make a wine that is accessible and fun and not expensive. He loves that kind of wine—basic, fun jug wine. There's nothing cynical about this; he just loves the kind of wine that he grew up on, wine that his family would enjoy for everyday drinking. So no, Rosso or Bianco are not wines that are attached to his stewardship of this historic property. They're wines that are attached to his life."

Rosso and Bianco, in addition to bringing the Coppolas great pleasure, have become wildly popular and tremendously successful in the marketplace. They are well-made blended wines with memorable theatrical labels connected to a celebrity of cinema, selling for about $10. So with this segment of the market covered, as well as the mid-range (Family Wines and Pennino), and high end (Cask and Rubicon), one might think that Eleanor and Francis Coppola can take a moment to savor their hard-fought success in the American wine industry. That will never happen, according to Eleanor.

"Francis is a dreamer, but Francis is basically a builder. As soon as something is built, like Rubicon or Rosso, he looks around and asks, 'OK, what else can we build?' I see him taking so much pleasure in finishing the rebuilding and refurbishing at the Inglenook château, and part of me is wondering, anticipating, 'What's next? It's built now, so what's next, Francis?'"

That's a question that the employees of Niebaum-Coppola might tend to shy away from, because they know two things about asking that question. The first thing they know is that Francis will have an answer, and the answer will not take the form of a modest proposal. The answer will be another Grand Design. The second thing they know, and this is the frightening thing, is that the Grand Design will be exciting and challenging, and they will get sucked in by that excitement, that challenge, even as they shake their heads and mutter to themselves.

Most Grand Designers leave the details to other people who are good at being able to realize pieces of the Grand Design, but have a hard time seeing the whole picture. According to Eleanor Coppola, no detail is too small to be addressed, and re-addressed.

"Francis and I believe that there is no better fertilizer for this property than the owner's footprint, so we sit out by the château and observe things that we'd like to tune up a bit, things we want to change, things we just like to enjoy.

"Francis is keenly aware that visitors to the property seem to appreciate that he's here; people do want to see him. He goes over to the château and sits outside in the park, quietly sipping an espresso or a glass of wine, because he created that park and re-created the château, and he likes to go over and look at it. He's really into it. He just didn't send somebody out to make this business."

Much like his predecessors, Gustave Niebaum and John Daniel, Francis Coppola possesses the characteristics that come naturally to a person with a vision who is channeling and combining that vision with far-flung entrepreneurship. He is an exciting person to work with because of his entrepreneurial vision, but along with that excitement come such exacting standards that they border on compulsion or fixation.

"He drives everybody who works here crazy, because he wants all the details to be right. Because of his film background he is so into the smallest details. He sees something and feels compelled to move it three inches to the right or left. And when the details are seemingly satisfied, he's got a whole bunch beyond that. Of course, he always needs a producer/winemaker/manager; he gives the idea but somebody else has to carry it out. Francis addresses the big things in life. So, it's exciting to work here, but you have to deal with Francis' demanding temperament. Some people get it, and a few people don't."

"The same obsessions that work so well in Francis' best movies translate to both the property and the wine. I have a lot of regard for it, but it's not me. It's also a male-female thing. As a mother there are so many times you have to adjust and compromise. Your kid is sick, so you have to change your whole life, your daily plan. That's my role. I'm working on the smaller, almost unseen parts."

Eleanor Coppola indulges and supports and applauds her husband's unique approach to life, but knows who she is and what she needs. She requires much more personal time and space than does Francis, and it is important to Eleanor to create that space, both for herself and for her family.

Since 1975, the Coppola family has lived in Captain Niebaum's house, which was also the Daniel family residence. It is an unpretentious, large Victorian house, not the showplace of a world-renowned film director. It is a family sanctuary, filled with photos, paintings, and mementoes, especially of Francis and Eleanor's three children. Above the fireplace is a painting of Francis and Eleanor's eldest child, Giovanni "Gio" Coppola, who died in a boating accident when he was twenty-three years old. Above the TV set and stereo is a striking Richard Avedon photo of their daughter, Sofia. Stuff from younger son Roman Coppola's childhood and adolescence. A little drawing made by the Coppola's granddaughter, Gia. It's cozy and a little funky, and that's the way Eleanor Coppola likes it.

"The house is personal; I draw the line at the front door. I was approached by all these antique dealers, who wanted to make it into a movie set. I live around sets enough to know that I didn't want that. I didn't want the interior done to *House Beautiful* standards.

"When we moved here we had three kids, dogs, and I didn't want to live in a museum where I had to worry about the kids spilling a coke on the fabulous antique thing. What you sit on is comfortable and serviceable.

"I felt the house was a work of art, so I didn't change the ceilings, the floors, or the walls. In fact, the only change I made was putting a window in the kitchen.

"If my kids made something, a tsotchke in kindergarten, I wanted to be able to show it off in comfortable family surroundings. It's really where we live."

If Eleanor Coppola is protective of her family's home and their right to privacy, she is just as aware of her role as the steward of the extraordinary property that her family owns, and the responsibilities and obligations that come with that stewardship.

"I feel, since I find myself in this unexpected position as the owner of the property, as the conservator of this property, that I want to take care of it, so that when I die it's as good or better as when I found it. Hopefully better. I really am just a steward; this piece of property is too incredible to be one person's possession. I don't mean to sound overly precious, but I really do believe that in a very real sense it belongs to all of us. It is my responsibility to take care of it, and my efforts are towards that."

As she talks about her present-day and future responsibilities, Eleanor Coppola, who is often reserved, becomes passionate, even poetic, about the gardens around the family residence. The gardens take on an overtone of metaphor as Eleanor articulates her role in their care.

"This garden is orchestrated; there's always something blooming. First, the dogwoods, and as they fade, the crabapples, and as they fade out the iris begins to bloom. It's just like an orchestra the way it was planned. Look. We're sitting under this extraordinary Japanese maple!

"This place is my domain. Francis likes it, but it's not a focus for him. The garden was planted in 1890, and the main trees, the bones of the garden, are over 100 years old. I care that we've thoughtfully planned for the future, with a lot of big trees going into that future. I want to maintain this so that someone won't come along after I'm gone and say, 'Oh, my God, they ran this place into the ground.' That goes for everything, especially the soil. I just want to do whatever I can to maintain this land, and to bring it to its full potential."

Eleanor Coppola is acutely aware of the fragile ecology of the estate, and realizes that to keep the property thriving and in balance, she, not entirely unlike her husband, Francis, must tend to details in a big way.

"There's always something wearing away, and it's not the glamorous part, it's the boring part. I feel I have to look out for that part that keeps things from deteriorating. Francis makes the show and builds the vision, but I feel compelled to take responsibility for keeping everything alive and healthy.

"I'm the person who maintains the property, maintains the history. This is a profound treasure, so it always astounds me to find myself here.

I'm never blasé about the property. I have to ask myself what twist of fate brought me here? I know it's not my doing."

Eleanor Coppola is talking about twists of fate, turns of destiny, as she remembers how her family came to own and care for the most historically significant vineyard in Napa Valley.

"In 1975 we came looking for a house in the Napa Valley, a cottage with perhaps a few acres to grow some grapes to make a little wine for the family and some friends. The real estate agent offered to show us this place, the Daniel residence, 110 acres of vineyards, and more than 1,500 acres butting up against Mt. St. John, just so we could see it. We never thought it would be anything we would be interested in buying.

"To me it looked like a movie set, it was beyond anything we could imagine. I thought the property was completely beautiful, breathtaking, but I grew up in a small family with a small house, small everything. Francis can take on larger projects and not be intimidated. He's kind of operatic and can see the big picture, and then follow through on it.

"The property we looked at was the Daniel family estate. The house was built by Gustave Niebaum. In 1971 John Daniel died, and his widow, who was a Mormon and against making wine, put the house on the auction block. It was bid on by a number of people, including us, and was acquired by a group of twelve investors, who called themselves Oakville Associates. They wanted to develop the land on the hill, and sell off the house and vineyards for what they paid for it. They essentially wanted to get the mountain for free.

"Oakville Associates had drawings for sixty homes situated around the meadow, but in 1974 this part of the Napa Valley was declared an agricultural preserve and they couldn't proceed with their development plans. They were discouraged and wanted to get out.

"So, Oakville Associates knew we had bid on the property and they approached us and several other bidders. They came and asked us, and we stepped forward and said yes, we would buy them out. Francis had made enough money on *Godfather II* for us to buy it. In 1975, we paid in the vicinity of $2.2 million for 1,700 acres, including 110 acres of the Inglenook vineyards. So, we owned it. It was kind of overwhelming at first."

How is it that the Coppolas, who knew next to nothing about wine and the wine business, how is it that they were able to buy this incredible land? This was a big part of Inglenook, and now it was owned by people who loved it, but didn't quite know what to do with it. Why didn't Robert Mondavi buy it? The Gallo brothers?

Eleanor Coppola can shed some light on the question.

"At that time—1975—vineyard property was about $7,000 per acre, so for 110 acres of vineyard, you didn't want to pay $2.2 million. A person looking for vineyards would not choose this property. I am sure everybody thought we overpaid, but twenty-five years later it looks like a bargain."

From the day they bought the Daniel estate, Francis and Eleanor knew that one day they would have to buy the other part of Inglenook, the front property, including the château and less than 100 acres.

"The property in the front, the vineyards and the château, were owned by Heublein, who owned the Inglenook brand. We were acutely aware of what Heublein was doing. They were ruining the property. They wanted a parking lot all the way back to Niebaum Lane. And they built that barrel building which completely obscured the view of the château from Highway 29. Heublein wanted to build more buildings, make it an industrial park. When Francis came back from making *Apocalypse Now* he had a plan for that storage building. He wanted to blow it up."

Not long after purchasing the Daniel estate, the Coppolas' fortunes began to nose-dive, bottoming out with a bankruptcy caused by *Apocalypse Now* and Francis' next film, *One From the Heart.*

"We had big debts. Francis insisted on paying the bank more than $16 million for the *Apocalypse* negative, since he felt it was the honorable thing to do. It was a tough time. Francis had to take whatever directing jobs he could get that paid him enough to pay the bank. We almost lost this place more than once. On one occasion we were saved from the auction block just two hours before the auction was supposed to begin. All the time Francis kept talking about buying the other piece of Inglenook. He strongly believed they belonged together and that they should never have been split. It was all about history for Francis.

"One day, I was out here in the garden and it was one of those dark moments where we might lose this place, and I had one of those realizations that everything was going to be OK. Francis seemed to be undaunted by the scope of the whole thing and I guess I felt courage in seeing his courage. I thought, 'I guess if he feels it's OK, then it's going to be OK.' And here we are at the twentieth harvest."

As it turned out, owning the *Apocalypse Now* negative was very profitable for Coppola; it was and continues to be a successful picture. He made *Bram Stoker's Dracula,* which was quite successful, and the Coppolas could begin to see light in the money tunnel. And he continued to talk about buying the rest of Inglenook, reuniting the property. He even sent the chairman of Heublein, Stuart Watson, an unsolicited check for $10,000 as a down payment on the property. Watson was not amused. Eventually, however, in late 1994 the property did come up for sale.

Now the Coppolas, twenty years after acquiring the Daniel estate, were on the precipice of acquiring the rest of Inglenook. Francis and Eleanor Coppola had witnessed, on an almost daily basis, how a corporate owner, with its offices and decision-makers in Connecticut, had let a jewel revert to coal. By the time Inglenook was put on the block, Heublein had stopped making wine not only at the estate, but also in the entire Napa Valley, and was using the Inglenook brand as its label for inexpensive wines produced in the Central Valley.

The Coppolas realized they would have to pay a lot of money for the right to claim Inglenook. Eleanor remembers that "for the new property we knew we would have other bidders, and that made the price escalate dramatically. What was appealing about the property was that it is an historical vineyard, with the château, with access to the highway, and a winery permit, which is worth a lot. So, other bidders came to the table, but Francis made an intentionally high offer, and eventually we were able to reunite the properties."

So, since 1995 the Coppola family has owned what once was Inglenook, including the château—once the site of the winery, and now the site of a public space—dedicated to celebrating Francis' focus of family, film, food, wine, and a complementary lifestyle. The tasting rooms and sales from wine, packaged food, books, t-shirts and sweatshirts, videocassettes, cigars, and upscale souvenirs generate in excess of $100,000 per month in revenue on a year-round basis, and as much as $400,000 per month in the summer and fall. This helps in a significant way to put the family wine business, which now has seventy-five employees, on a firm financial footing.

A factoid you hear thrown around the Napa Valley is that although the Napa wine country is the second-most popular tourist attraction in California (eclipsed only by Disneyland and just edging out the San Diego Zoo), the average tourist visits only three wineries. Of the three, two are givens: Robert Mondavi and Niebaum-Coppola. This means that tens of thousands of visitors pass through the portals of the Inglenook château, and almost all of them buy something, if only a taste of Rubicon, or a poster, or a bottle of Bianco.

Eleanor Coppola, who seems like the last person to enjoy a steady flow of big crowds, has a unique take on Niebaum-Coppola as a tourist destination.

"Gustave Niebaum tried to re-create what he saw in Europe and copy what he liked best about the European lifestyle, and the château is part of that sensibility. I think of it as an historic monument, an important building, and I believe it's a gift to be able to share it with the public, invite them in, let them see a remarkable example of period architecture. It's someone's dream of what a great château and vineyard in the Napa Valley could be, but modeled after the European vision.

"Francis creates the celebrity buzz, which is an element of drawing people in, and then when they visit, I hope they have a sense of authenticity and quality. When you step into the space you don't think it is just another gift shop, but it's an extension of our family's lifestyle."

"If, for example, you buy our pasta, and take it home and cook it, you should enjoy it, and not think of it as some tourist rip-off item. My goal would be to hear from the people who really enjoy our products. We hope we can do this and still be businesslike about it, so the winery can support itself."

Eleanor Coppola knows that being an integral part of a large and highly visible family business is not achieved without personal sacrifice, especially as she and Francis approach a time in their lives when they both would like more time for reflection, more time to themselves. Living in the public shadow of a household-name husband who has received five Academy Awards, it is easy to forget that Eleanor Coppola won an Emmy for her documentary film, *Hearts of Darkness: A Filmmaker's Apocalypse.* She is working on a video documentary about her friend, Alice Waters, the famed chef/proprietor of Chez Panisse, and a leading voice for organic agriculture and sustainable communities.

"I've known Alice Waters since she was a Montessori School teacher. Alice has made me much more aware of so many things. I traveled with her to the Land Institute in Kansas and learned a lot more about her philosophy. She's a revolutionary, and we are both dedicated to getting her message out. It's hard, almost impossible, to make a documentary and be active on an almost daily basis in our wine business."

Eleanor Coppola is a strong and supportive person, but she does not fit the portrait of a tireless cheerleader for Niebaum-Coppola and the Napa Valley. She is too sincere, too thoughtful a person to play that role. Still, she is beginning to see the family business grow in a direction that pleases her and gives her a grounded sense of hope for the future of the property and the business.

"I've had my moments of regret. I wanted more time to write, shoot video, paint, and other personal things. There are endless responsibilities and interruptions that go along with this property. A tree falls in the road, you can't leave it there.

"Now the business is a larger organization and has stronger management, which is a very positive step. We're getting closer and closer to having the right people in the right place. As the winery has gotten bigger, we can support top-flight professionals, like Scott McLeod, our winemaker, and Erle Martin, our general manager.

"A small business really necessitates the owner being there every minute of every day. As a small business, we were maintaining, but not developing, the brand of Niebaum-Coppola. As we've expanded I can see that we can attain our goal of getting the quality of the wine and its image up to where it should be.

"We always knew that the property would take a very rich person to keep it going, so our goal is to have the winery evolve to the point where the winery can support the property, so that it won't get sold off in pieces for payment of taxes by our kids, our heirs."

In the United States it is the rule, not the exception, for first-generation family businesses to wither in the second generation, and die in the third generation. Will Sofia and Roman Coppola want to continue the stewardship of the property, continue to honor the history and tradition of Inglenook in the form of Niebaum-Coppola wines? The question is an important one to Eleanor Coppola, and she is hopeful.

"I think my children are at a stage in their lives where it's really important to get away from home, away from the family, and develop themselves. Sofia is a clothes designer (her partner is a young woman with whom she attended high school in St. Helena). Her clothing company, MilkFED, is based in Japan, where she's opening a store. She's traveling a lot, and she had a little part in the new *Star Wars*, just to wear the costumes of a handmaiden.

"She just directed her first independent 15-minute film. Her photographs are in exhibitions and magazines. Sofia comes home to get centered, focused, and grounded. She and I and my granddaughter, Gia, went to the lake to skinny-dip and to lay in the sun yesterday. I feel honored that instead of running off to a spa with a friend she chooses to come here.

"Roman has made more than thirty music videos and was nominated for a Grammy. Music videos are a great form; a short little mini-movie, a great place for young filmmakers to get started. He's been directing commercials for Honda and Levi's, and he shot the second unit for Francis on *Rainmaker*.

"I think this place, this land, this house, is in their blood; it's their home, it's where they grew up. I hope as they grow older and have their own families and bring their kids here, that they will just end up here."

Eleanor smiles as she adds an emphatic coda to her belief that the Coppola children are destined to continue as the conservators of this very special place.

"Both of our kids have the same earth sign — Taurus — and I figure sooner or later that's got to kick in."

9
Consulting with the Captain

Chatting with Francis Ford Coppola is a bit like shooting the breeze with one of the guys from the old neighborhood who became worldly and wildly successful, and returned to his old stomping grounds, but just for a short visit. His conversational language and attitude is approachable. Like so many of the best writers, he has a gift for simplifying complex ideas. And his ideas *are* complex—big and complex. And so is lunch.

Before we talk, we eat, and before we eat, Francis cooks. In the large kitchen of the Coppola family's Rutherford home, dressed in his usual khaki shorts, bright yellow Hawaiian shirt, and wearing sandals, he is busy making *spaghetti alla primavera*, a dish native to Basilicata, the province of Coppola's southern Italian roots.

"This a very fine pasta from a small Neapolitan producer; you can't get it in this country because it's not fortified with vitamins. The sauce is made from fresh tomatoes and fresh basil, but it's not completely cooked. The hot spaghetti warms it up."

While waiting for the water to boil for the pasta, Francis slices some homemade chorizo, lays out some prosciutto di Parma, tosses the salad, grabs the olive oil and balsamic vinegar, and cuts hunks of baguette. He moves gracefully in the kitchen, and is intensely focused on the task at hand. Attempts at small talk are reciprocated with nods, grunts, and distracted one- or two-word answers.

When I tell him that I went to the new movie theater in St. Helena (owned by the Wagner family, proprietors of Caymus winery) to see *Ulee's Gold*, and I expound on the brilliance of Peter Fonda's performance, Francis Ford Coppola, five-time Oscar-winner and lifetime student of the cinema, could not be less interested.

As the steam from the boiling pasta water fogs his glasses, he lets me know that "I don't like to go to the movies. I'll go back one day, but now I'm moving more towards a reclusive life, trying to figure out how to phase myself out, trying to figure out how I can enjoy everything I've created. Let's eat."

We carry the lunch, which is enough food for about eight people, out to the verandah, where a table has been set for us. With a sweeping view of the Gio vineyard glistening in the sun, we sit down to eat and to talk. Francis Coppola serves me a heaping bowl of spaghetti and sauce, as I open a bottle of the 1995 Coppola Family Cabernet Franc and pour for both of us.

"I love this wine for lunch, although I wish we could have the Pennino Zinfandel. But my wine guys tell me there's none left, so we can't get any."

Francis smiles as he makes this remark, but without a tinge of irony. He is happy that the Edizione Pennino Zinfandel, named for his maternal grandfather's music publishing business, and perhaps the finest Zinfandel produced in the Napa Valley, is enjoying such an enthusiastic public reception. The fact that he owns this estate, and has sunk between $25 and $30 million of his own money into it, and still can't get a particular bottle of his own wine, doesn't seem to faze him. He knows that there will always be plenty of wine for him to enjoy.

We raise our glasses, and Francis offers a toast to my family. I offer one to his. Immediately after the glasses clink, and just preceding the hearty, happy slurping sounds of eating and drinking, he brusquely breaks the spell of the moment by offering a criticism of his own cooking.

"The pasta. I thought I took it out early, but not early enough. I like it firmer." The spaghetti seems perfectly *al dente* to me, but as I learned over the course of several months at the Niebaum-Coppola Estate, God is in the details, and if God doesn't get there first, Francis Coppola will.

"I'm sort of wired up in a funny way. I've always been a very spatial person so I don't have any difficulty doing many things at once. It's just my nature. I don't ever do one thing at a time; I'm not capable of it."

Francis Ford Coppola is talking about himself, trying to explain why he needs to write, direct, and produce movies, publish a magazine (the literary journal, *Zoetrope*), own an eco-resort on Belize, and produce wine from some of the most historically important vineyards in California, and then obsess for twenty years until he could buy the adjoining property, thereby reuniting the original Inglenook, its land, buildings, and massive stone château.

Even before Francis attempts to explain himself, it is clear to the most casual observer that he is a person who entwines his many passions, labors, and interests. You need only visit the old barn where his estate-bottled wines are made for his personal cult of multitasking to make an indelible impression.

On the first floor of the yellow barn is the winery, and it is tiny. Hardly state-of-the-art, the facility is the province of winemaker Scott McLeod and his team of assistants. Everyone at Niebaum-Coppola acknowledges that a new winery and storage facilities are necessary, but they also realize that Francis has a Big Idea for winemaking operations. He wants to produce wine in the château, as Captain Niebaum did, as John Daniel did. This will be a neat trick, as the Napa Valley is Permit Hell, and many layers of bureaucracy will need to bend to Francis' will if the winery is to move to the historic building. So, for now, the little winery sits in the barn close by the Coppola family home, occupying only the first floor, as the extraordinary grapes brought in from the estate coupled with Scott McLeod's passion, patience, and practice produce stunning wines.

On the second floor of the yellow barn is Francis Coppola's personal library. The collection of thousands of books, periodicals, videotapes, films, slides, photographs, scripts, documents, and posters take up the entire floor of the barn. The collection is extraordinary, not only for the amount of material on the library shelves, but also for its interdisciplinary diversity and depth. A full-time archivist is employed for the library, along with several research consultants.

On the third floor of the same building is a digital audio recording/ editing facility (the sound for Coppola's film, *Rainmaker,* was being edited when I was there). The studio is crammed into the eaves of the barn, and it has its own third-floor entrance where a lot of people with cellular phones come and go.

On several occasions, winemaking has had to stop downstairs, and everyone had to be quiet (not a stretch for those in the library) because some live sound was being recorded upstairs. Nobody who works for Francis finds the all-in-one winery/library/digital audio studio at all unusual. As Scott McLeod replied so many times to questions about the perceived quirks of his employer that it became his good-natured mantra, "That's just Francis."

A common sight: Francis Coppola takes a leisurely stroll, walking from his house to the barn, smoking a cigar. He walks into the barn, not saying much to anyone. A little while later, he takes the same walk, but only in reverse. He is carrying a book or two. If he doesn't feel like walking, he drives back and forth in his classic beige Citroën 2CV/Deux Chevaux, the utilitarian, unpretentious car made popular by French college students of the 1950s and 1960s. Coppola's 2CV is one of only 2,000 in the United States.

Clearly, Francis needs to be involved in many projects simultaneously, and on a large scale. He seems to be chasing and is chased by dreams that other people would not even let themselves imagine. But how is it that he came to be the owner of these vineyards, this land, this place? How do you dream that?

Marshall McLuhan once wrote that "Hollywood movies are the dreams that money can buy." Is it that Francis Coppola was conditioned to such opulent dreaming that becoming the owner of Niebaum's property and the savior of the heritage and history of Inglenook was second nature to him? Why Coppola, the filmmaker, the dream weaver? Why not Mondavi or Gallo or Sutter Home or Beringer, successful winemakers with their feet set in the soil?

"I would question if money alone could buy this place. I came into possession of this place by accident. A place like this is not about having money. You have to be lucky, and you have to have a bit of a vision. Traditionally, people who owned a place like this wouldn't want to sell it at any price.

"But in 1975, when we bought the original property and house, it loomed as a bit of a white elephant. Everyone in those days were afraid of maintenance and upkeep, and today, of course, this property has become extremely valuable. Always the phrase back then was 'white elephant,' but I don't think it was a daring financial transaction even in those days.

"The reason we got both properties was not our money, but our passion. We've been approached several times, especially recently, with the idea of selling a piece of the estate or all of it. And it would be a lot of money, but what would we do with it? If we had a lot of money I don't know what we would do with it, beyond wanting to live here. I mean, you could be Bill Gates and you couldn't buy this place. I have no motivation to sell it."

But how is it that the Coppola family was able to buy the most historically significant vineyards in the Napa Valley, not once, but twice? Why, in 1995, twenty years after the Coppolas bought the original piece, and the California wine industry was beginning to shine, why were they able to buy the front property and the Inglenook château? Surely the family and corporate vignerons of the Napa Valley wanted to own a piece of valley history. This is a question that engages Francis Coppola's personal world view.

"I am a very big believer that there are things in the makeup of the universe that we don't understand, and so it builds a lot of room for such phenomena as fate and intuition and things that are hard to explain. But beyond that, a property like this would only become available to so-called 'rich people,' and I've noticed that rich people have a very ironic habit of trying to figure out how to prevent their last dollar from getting away, rather than focusing on the hundreds of millions of dollars that they make. By nature, they're cautious, stingy people, and I think that's why rich people lost the opportunity to buy and own this property.

"There were a number of wealthy people, some of them very well known, who considered buying this property, but they lowballed it. They just couldn't stand the idea of just coming out and buying it no matter what the price was.

"So, all the wealthy and wine-interest people made lowball offers. We made a bid and lost it, and were able to buy it on the rebound. So, when we bought it, it wasn't a matter of price; we just wanted to buy it. It wasn't in my mind to negotiate; it was just to get it."

If this sounds like the voice of someone who is not a good businessperson, someone who believes that with enough money you can buy anything, even if it doesn't make sense on the balance sheet, you may not be really listening. Coppola is passionate about this land, and as far as the bottom line, he is far smarter than someone who is just throwing movie money around because he wants what he wants.

Confirmation of Francis Coppola's view comes from Dennis Fife, owner of Fife Vineyards in Napa and Konrad Winery in Mendocino. He had worked in Heublein's fine wine division starting in 1974, and was president of Inglenook from 1984 to 1989. Heublein owned the 100 acres of Inglenook vineyard fronting Highway 29 and the Niebaum-built château, as well as the Inglenook brand name. They also owned (and still own) the Napa Valley's other historic jewel, Beaulieu Vineyards.

According to Fife, "Coppola got the first piece, the Niebaum/Daniel estate, in 1975, because Heublein thought they were going to get a good deal, and we offered a ridiculously low bid. It was particularly stupid because that piece of property was very important to them; they could own all of Inglenook. *And it was a matter of somewhere between ten and fifty thousand dollars on a multimillion dollar piece of land that made them lose it.* Really stupid."

Francis and Eleanor Coppola bought that property, about 1,500 acres, including 110 acres of Inglenook vineyards, for approximately $2.2 million. In 1995, they bought the rest of Inglenook—including 110 acres of vineyards and the stone château—from Heublein for about $11 million. Although the Coppolas knew that they were destined to buy the property from Heublein, negotiations were rocky.

"The sellers knew we were willing to pay their price, so they tried to take advantage of us. They kept coming back to the deal, trying to build in things, like 'We won't sell you the vineyards, just the château'! It was outrageous.

"This is still an issue in my mind, because I feel they were unfair with me as owners, and now I'm the owner, but part of the deal is I have to sell grapes to them until the year 2000, which I'm not happy about.

"The reaction by some people in the Valley when we started making wine seriously was that we were crazy, and when we began to restore the château, some people used to call this 'Francis' Folly.' Can you believe it?"

When things were tough at Niebaum-Coppola, after Francis took on more than $16 million in debt to Chase Manhattan Bank in 1980 to bail out *Apocalypse Now,* many of the cognoscenti of the Napa wine industry believed that Francis might fail, and ironically he was allowed to pursue his dream of a Grand American Wine without interference and without community support.

"I think among some of the people who live in the Napa Valley, I'm still considered a Hollywood filmmaker and a celebrity, even though this has been our home since 1975, and our kids went to school with their kids. And because I paid a lot to reunite Inglenook, I'm considered a patsy by some people.

"I am startled to feel that our family would be treated with anything less than the same goodwill we were years ago. I think there were rumors that this place would become available, because I zigged when I should have zagged financially. But lately, we're getting a little more power with the media, who know that we have *the* place, and our reputation may be coming into conflict with their expectations.

"Now that we've emerged from a boutique winery to a real château estate in America, and these same people see that we have the most historically significant wine estate in America, and they see that we have this beautiful and popular public place, and we're a real company, I feel there's a bit of dirty pool with me by the regulators, the people in power. They're tougher on us than they are with themselves. It's like, 'Hey, he gets all the publicity.' And although some people, like Robert Mondavi, are as helpful and gracious and supportive as they ever were, right from the beginning, I guess other people are worried that the boob has become formidable."

Coppola's complaints seem to have some merit. He must go through a tremendous amount of permit-seeking and board approvals to restore the estate to what it was in the era of John Daniel. Napa County seems intent on protecting the alterations of the Heublein era, which include the hideous storage building that blocks the view of the majestic Inglenook château from Highway 29. Perhaps even more important, Coppola will have to jump through many hoops to bring winemaking back to the château, because Heublein stopped producing wine altogether at Inglenook in the mid-1970s.

"The winemaking team is so passionate, and I want them to have a beautiful facility. I want drawings, and I want to know how much, and then we'll have to pay for it as a capital outlay. That will be the last frontier of the winery. It's good that we built the visitor's center first, because it shows the strong interest and demand, and so will benefit the winemaking team to get what they want."

Clearly, Francis has decided that he will have his winery, and he knows where that winery must be located, because the history and heritage of Inglenook dictate that the winery be sited in a specific place.

"We are building a winery, and it will be built at the Inglenook château. Rubicon is our wine that is all about this property, this place. I insist that we make Rubicon in the historic Inglenook winemaking facilities, in the château, and I'm told that we don't have the right permits, and again I say, 'Guys, don't hold me by what Heublein did here. They didn't care about making wine here.

"How can the county not let us make wine in the Inglenook winery? It feels like they have their hands around my throat, holding us to the state of affairs that existed when Heublein shut down the winery. Legally, we're helpless if they tell us that we can't make wine in that building.

"They're saying we're a new winery because we bought this from

Heublein in 1995, and as far as the county is concerned there was nothing here before. They're even telling us that Heublein was doing certain things without the right permits, and now that becomes my problem.

"It's all politics in a way. The people in power have been discouraging us from restoring and improving the property since day one, and I still don't know why. They just say, 'Oh, it's the rules.' And we have to be cautious because we don't want to make application for something and be turned down."

Getting a permit to build a new winery in Napa County is a bureaucratic nightmare. Building a new winery that conducts public tours and tastings, sells wine-related merchandise, and serves food is a virtual legal impossibility. Francis Coppola argues that what he proposes is not a new winery at all, but a return to winemaking at Inglenook in the original building to fulfill the original promise of Captain Niebaum and John Daniel — to produce America's finest wines.

"Gustave Niebaum and John Daniel made wine here for almost one hundred years. Let's face it, the county stood by and watched as this place, Inglenook, was destroyed. Now they want to treat us as though we're nothing more than Heublein's replacement, and I'm saying, 'Hey, we're not Heublein, we're the original Inglenook.' I would hope that we'd be treated like Inglenook, not a post-'90s Heublein.

"We're good for the valley, we're good for heritage, we're good for conservation. Yet, I'm beginning to feel there's a sentiment that I'm not a celebrity boob after all, and that we are serious about wine, and that we might disturb someone else's financial situation.

"I'm all for no-growth policy, but I think there should be more policy concerning heritage. Right now, there's a lot of financial incentive for breaking things up, selling off pieces, taking companies public. There's no incentive for people who want to keep the original heritage, preserve a family-owned business. Why isn't the county's position to encourage preservation? I would never ask for liberal growth laws, but I would like to see intelligent heritage laws. There should be incentives not to sell to investment bankers.

"What I would like the county to say is 'OK, Francis, you can be as Inglenook would have been had it not been divided.' Then we have to look at what Inglenook would have become. That division impacted it greatly. It's like a kid has a terrible auto accident when he's nine; for years he can't walk or talk. Then, a miracle operation comes along and he's thriving. So, I would like the opportunity to see this place blossom consistent with its heritage. I don't want to mall-ize it.

"There are far more wealthy people than me in the Valley. The fact is we're a 100 percent family-owned wine estate with very little debt, and for that to work, I've got to focus all my attention here. A lot of other people have portfolios, investments, etc., but I'm just willing to risk it all here. And now that the château is finished, people really like it, and all of a sudden they realize here was a vision, and here is a vision for the future. So now, we have public support.

"I have to be patient. I have to be optimistic. I think that policy will become more enlightened over the next several years, and I think, I hope, that lawmakers will see that with the act of preservation of the vineyards, our uniting of the Inglenook Estate, we should be allowed to have traditional rights of usage. I think our view will prevail."

Jumping back from the future to the present and the not-so-distant past, Francis Coppola is not quite so sanguine, and his tone is not so optimistic or hopeful. He wants answers to some important questions.

"Isn't the county, by doing things like protecting that storage building out front, trying to assure that we don't leave the Heublein era? Why? Heublein had carte blanche. Where was the county then?"

The creator of the three *Godfather* films pauses, then delivers a piercing blow to those who do not appreciate his dreams, labors, goals, historical perspective, and overall vision for Niebaum-Coppola.

"It's like a little Mafia here, and that's annoying."

"I came close to losing this place."

Bankruptcy is a subject Francis Coppola doesn't like to discuss for several reasons, and chief among them is that, as a story in the media, the story has been overreported and reported inaccurately. Also, because it is the story of a highly visible director, writer, and producer, who is the head of Zoetrope, his own production company, the bankruptcy has been reported as an integral part of Coppola's body of work, almost as *Bankruptcy: The Movie.*

As Coppola talks about his financial difficulties, which just about consumed his life from the years 1980 to 1990, his annoyance and exasperation turns reflective. He attempts to report the facts of the situation, and the toll it took on his life and career, with surprising equanimity, perhaps kindled by the luxury of hindsight.

"You know, people talk about and write about me going bankrupt several times, which is not really true. It was always the same bankruptcy, actually a Chapter 11 reorganization, all related to the same thing, paying the Chase Manhattan Bank more than $16 million to bail out the negative of *Apocalypse Now.*"

"If I wanted to, I could have just gone bankrupt, and that would have wiped my debt out. I would have lost the negative of *Apocalypse Now,* which ironically has made about $125 million over the years, and continues to make money, but I would have avoided paying back Chase Manhattan.

"This is our home, it's where our kids were raised, and it was very important to me, even when bankers were literally showing up here demanding that I pay this debt, that we not lose it. George Lucas is a wonderful friend to me and to my family, and he was totally ready to buy this place and hold it for me. But I made a choice to do what I thought was the honorable thing, to pay back this debt to the bank.

"When you get involved with doctors or the law or banks or any of those kinds of institutions it's easy to get in, but not so easy to get out. If I had just gone bankrupt, I could have saved a lot of money and I probably could have saved my career in a way, because one of the reasons I got involved in directing so many pictures, regardless of whether or not they were my scripts, was because I had to make a bank payment every year. I had to make a movie a year to pay off the bank."

Francis Coppola is nearly sixty years old, and as he continues to talk about the toll that these events extracted from him, it is clear that this bankruptcy was not just about money. The money, the overwhelming debt, the financial obligation, seems almost beside the point. Francis did what he thought was the honorable thing; he paid his debt. He maintained his honor, but, as he now confides, lost so much more.

"This came at a time when I was creatively my most potent. I had made *Godfather I, The Conversation, Godfather II,* and *Apocalypse Now,* all in a row, and that's where I was at age 40.

"The prime of any man's life is age 40 to 50, and this bankruptcy, paying the bank, just totally took that away from me. I've often thought that had I known how complicated and painful it all was, and how when you do the honorable thing and pay your debt back they still penalize you in ways you can't even imagine, I just would have gone bankrupt. I would still have this house and the property because George (Lucas) would have bought it and held it for us. I would not have had to do ten years of those movies. I had to do *Peggy Sue Got Married* or whatever I could get that would pay me enough to pay off Chase Manhattan.

"Time is more valuable than money. I would rather have that decade back than any amount of money. Obviously, the only healthy attitude is that I don't regret my decision. I did what I thought was the honorable thing, and in the end I got *Apocalypse Now* and the bank didn't."

For the first and only time in our conversations, Coppola begins to talk just a bit about his films, but as I listen closely to what he is saying I realize that he is talking about a lot more than a particular movie. Perhaps he is talking about the thin lines between art and life and their fragile intersections.

"*Apocalypse Now* made money and still makes money. I think over the years that it is *the* movie about Vietnam. There have been many, but I think it caught the imagery and madness that Vietnam was, partly because we allowed ourselves to go to that place."

In *Apocalypse Now,* Francis Coppola created a singular vision for what is inarguably an historic and grand cinematic epic. The vision led to the pain of bankruptcy and an unexpected dramatic climax only midway through his career. Coppola also holds a singular vision for Niebaum-Coppola wines, produced at what is inarguably an historic and grand estate. Has he learned how to hold to his vision for Niebaum-Coppola and avoid the pitfalls that he experienced firsthand creating his vision for *Apocalypse Now?*

Francis Coppola definitely has learned from the experience, and part of what he has learned is so profound that he expresses it on the biological level.

"I learned to beware of people bearing money. You have to know up front that investment bankers are in it to end up owning the product. You have to realize that and become like one of those creatures in nature that gives a part of its body and hopes that the predator is satisfied with that."

Francis Coppola is explaining the source of his inspiration, the guiding spirit that sustains his own vision at Niebaum-Coppola.

"Whenever I'm going to do something here, a major decision with lasting significance, I ask myself 'What would Captain Niebaum have done?' That's why, like the great châteaux of Bordeaux, like Lafite-Rothschild, for example, we put the Niebaum name first, the Coppola name second. We honor his vision and continue that vision."

One example of Coppola's psychic consultations with Niebaum centers on the extraordinarily dramatic staircase that was built in the Inglenook château to create a vertical interior flow from the first to the second floor of the stone building, which is close to 120 years old. The staircase is made from beautiful woods from Belize, where Coppola owns a small resort. Francis pondered what kind of staircase Gustave Niebaum would have appreciated.

"A stone staircase? A wood staircase? If you look at the woodwork, the decorative elements in the château, I think the Captain would have liked a wood staircase. So we tried to build an exceptional staircase."

Another example of how Gustave Niebaum guided Francis Coppola's decision-making is illustrated by an important choice made in the Niebaum-Coppola vineyards.

"There came a time when we were advised to replant our vines, but on AXR rootstock, because of slightly better yields in the vineyards, and because they were supposed to be resistant to phylloxera. We were also told to pull out our Zinfandel vines. Would Captain Niebaum put in AXR just because it yields a bit more? No, he wouldn't. And something told me not to go along with the AXR phylloxera research, and that turned out to be a wise choice, because AXR is riddled with phylloxera in Napa and

Sonoma. Those Zinfandel vines are the source of our Pennino Zinfandel, which is a very special wine. Again, I believe Niebaum planted that Zinfandel and he had a good reason to do so. We didn't pull it out."

Perhaps Coppola feels such a deep and abiding respect for the judgments of Niebaum because they both came to this property motivated by passion, not profit. Niebaum was not a vigneron by profession, and neither is Coppola. Niebaum had $10 million in 1880, the equivalent of hundreds of millions of dollars by today's standards, and could have bought a vineyard, or vineyards, anywhere in the world. Coppola, when he came to Inglenook in 1975, had no intention of becoming the padrone of this magnificent estate, but he, too, had the money to do so.

In looking at the obituaries of Gustave Niebaum, including one in the *St. Helena Star*, his ownership of Inglenook earned just a passing mention; his international business successes were highlighted. Indeed, Niebaum did not think of himself as a vigneron, but as a businessman with a passion for fine wine.

Not to put too ghoulish a point on it, but the same could be said about Francis Coppola. He is not a winemaker, but a filmmaker with a passion for fine wines. It is doubtful that Coppola's obituary will focus on Niebaum-Coppola, as, at least so far, his films define his claim to fame and recognition. On the other hand, Francis has said that he views "what we're doing here as part of my body of work." Whether the press and public will understand that point when assessing his life's work is another matter.

If Francis Coppola sees himself as the facilitator of Gustave Niebaum's legacy, and recognizes that he is not going to chuck filmmaking for winemaking, how does he see his role at Niebaum-Coppola, and has he cast an active role for himself in the American wine industry?

Coppola's answer is, not surprisingly, complex.

"I've been many times to Sacramento and testified and spoke for the wine industry, because I'm kind of a famous person, and I want to use that celebrity to help the industry. Here at the estate, I enjoy hearing about and tasting the wine, and putting my two cents in when it comes to ideas on, say, the design of a label.

"We've put all our money into this project, we didn't borrow it, we didn't leverage it. The transition from a boutique winery to a larger winery has been expensive and time-consuming. There were operations costs of maybe a million dollars to bring our accounting to a modern standard. That was greater than anticipated, because the winery was not accounting accurately; there were losses that the accounting department did not report. We were a year and a half off. So we had to bring accurate records to this newly expanded company, and it was expensive to reconcile the deficit. Now we have a terrific accounting team.

"I think I've made a tremendous commitment of all my financial resources to this project, and philosophically I think I've set the tone for the future. I think the road is clear, and that the winery will be able to support the family and keep the estate intact.

"I have a long-term vision for Niebaum-Coppola that extends beyond my stewardship, and that is still evolving, but I also know what I would like to see happen here five years from now. First, I want the entire infrastructure of the estate—the buildings, the gardens, trees, the water and electrical systems perfectly maintained and brought up to snuff.

"All of the vineyards are now planted, but we would like to develop another 30 acres of additional vineyard, which would be mostly mountain vineyards, planted perhaps with Italian and Rhône grape varieties.

"I would love to see the olive groves maintained and cared for, and ultimately make a little bit of estate olive oil.

"The vineyards perfectly maintained, producing 60 to 75,000 cases of estate-grown premium wine, with the winery sited in the Inglenook château, with the château being perfectly maintained and the museum fully realized and the library of wines maintained. We may lower the roof of the barrel storage, or build our own building to house a second modern winery.

"The bianco, rosso, and claret wines are sourced, and these programs might expand, but I would never want to produce more than 250,000 cases, 75,000 of which would be estate-bottled wines.

"I would like to see this home be the official place of residence and welcome. Eleanor and I might build a small cottage for us to live in, and use this beautiful home for receptions and special occasions."

Francis then moves his five-year plan to another level, initiating a conversation about the structure of Niebaum-Coppola as a company and a brand.

"In order to keep the property family-owned and to continue producing our 100% estate wines, we might look at spinning off another publicly held company, like 'Francis Coppola Presents' or the 'Niebaum-Coppola Company,' and that would protect this place from becoming that. I'm talking about a second branded company that in no way infringes on the estate or the estate wines. It might take in wine and foods, and branded stores, too.

"Any executives coming on board with us tell us scary things about how much money we need if we want to grow at the rate we're growing. They can find investment bankers who for one-third ownership in the company will give you that money. So it's tricky to navigate. The really smart executives who might join us are aware that if we go public it's good for them. They don't have a family stake in it, so they'll go on to the next thing. At the same time, our children, Roman and Sofia, are on the board and they do contribute their ideas. Their careers are quite dynamic, but they come to the board meetings and participate.

"Ultimately, for our family, it does come down to the same problem that the Mondavis have had. Mondavi is a public company now because the inheritance taxes that Robert Mondavi's children would have paid when he died would have been a nightmare.

"When my wife and I die, the government could tax away the estate unless I take measures now to make sure that doesn't happen. We carry life insurance policies that we thought would pay the inheritance taxes, but as the value of this place increases, it becomes prohibitive to carry such large policies.

"One day this place will be worth hundreds of millions of dollars. How will my kids come up with half of that in taxes?"

For someone whom so many people consider to be a dreamer or a movie director with megabucks, Francis Coppola seems to have a coherent plan for implementing his vision. The plan involves expanding the existing brands, creating and marketing new wines, and striving to make Rubicon better and better. Rubicon defines the *terroir* of the property.

When Francis Coppola introduced the 1978 Rubicon, which legally could be labeled Cabernet Sauvignon, because it is more than 75 percent of that varietal, with the remainder of the wine, usually about 15 to 20 percent made up of Merlot and Cabernet Franc, there was only one other Bordeaux-style blend produced in the Napa Valley: Insignia, produced by Joseph Phelps, who also produced a wide range of varietal-label wines. Rubicon was the only wine produced by Coppola, and most people thought he was misguided, if not crazy. However, Robert Mondavi, whom Coppola respects enormously, gave him encouragement, and the legendary André Tchelistcheff, who created Georges de Latour Private Reserve Cabernet Sauvignon for Beaulieu Vineyards, agreed to consult on the winemaking. This is consistent with the goals of Niebaum-Coppola; Francis reveres the history of the Napa Valley. Bob Mondavi is eighty-seven years old, and André Tchelistcheff, who began consulting for Coppola when he was seventy-seven years old, died in 1994, at ninety-two.

As Coppola maps his plans for an expansion of the winery and its brands, he leaves himself open to public comment and criticism. Unique to his position as a crossover public figure, evaluation of his wines and his plans are as likely to appear in *Esquire* as in the *Wine Spectator.* Having been a public figure for close to thirty years, he is, of course, aware of this scrutiny, but seemingly unfazed by it.

"My intention has always been that Rubicon be one of America's five or six greatest wines, but recently I came to the conclusion that there are at least 100 to 200 greatest American wines. So, my thinking has changed. Now I believe that we are a grand wine—by my definition, not only a great wine, but one with heritage, with the setting, with the vineyards in the epicenter of what determines a particular *terroir*. What I consider a grand wine has to be a wine that can please contemporary taste, but it has to have an historical aspect, and the vineyards must be at the zenith of what defines an area. Cabernet Sauvignon vineyards on this historic property in the Rutherford district of the Napa Valley make for a grand wine."

In May 1998, a second premium wine was released. "Cask Cabernet" is a 100 percent estate-bottled Cabernet Sauvignon. Three hundred cases of the excellent 1995 vintage were quietly released as a test balloon. The wine features the Coppola-obsessed, very expensive oak veneer label (about $1.50 each), and is luscious, delicious; a very fine Rutherford Cab, aged in American oak barrels.

"Rubicon is a singular wine, a grand wine, but looking into the heritage of the property we find that there has always been a Cask Cabernet, so we thought this wine would be more about California heritage, while Rubicon has its own heritage."

A Cabernet Sauvignon from the Inglenook vineyards is, by definition, a great wine, and a wine of historic interest. John Daniel, Jr. began to release cask-numbered and limited-cask Cabernets, and this helped to lift Inglenook's reputation to new heights. There is much anticipation at the winery and in the marketplace about Cask, and Coppola hopes that the wine will be received enthusiastically as the equivalent of a fine "second" wine from a Grand Cru Classé château of Bordeaux. Château Latour produces Les Forts de Latour. Château Margaux produces Le Pavillon de Château Margaux. Niebaum-Coppola produces Cask Cabernet as a "second" wine for Rubicon.

Coppola indulges his theatrical bent with a line of wines called "Francis Coppola Presents." The wines, *Bianco* and *Rosso,* a blended white and a blended red, are tasty, simple wines that are perfect accompaniments for a pizza or fried chicken, or any simple food. They are well made by Niebaum-Coppola winemaker Scott McLeod, but the grapes and juice are sourced throughout California, and the wine has no sense of place. It is just a pleasant, easy-drinking $10 wine, with an appealing label featuring bold colors and Coppola's signature. If Rubicon represents Inglenook heritage, and Cask Cabernet represents Rutherford heritage, then Bianco and Rosso represent Francis Coppola's personal heritage.

"Our family never drank premium wine. I never had any until I got it through movie people who could afford it, and I never knew wine could be like that. My father and my uncle made a little wine for the family and drank Gallo wines. I have great respect for the Gallo family and I think they're a great family. They popularized wines in America for decades and sold drinkable wines at pretty damn good prices. My family drank Gallo.

"Our Rosso and Bianco have been successful, and for everyday drinking it's what I like. It's so much like what my family drank—my original idea was to sell it in a bottle with a handle, an elegant jug, but the design was too difficult to execute—and I'm happy when I drink these wines, or maybe the Coppola Family Cabernet Franc.

"Rosso and Bianco came about because I always wished that in the little restaurants I go to in San Francisco's North Beach people would happily drink my wine, *con viviala.* I like to give my wine away, but I can't give

Rubicon away, because we're sold out. So, the Rosso and Bianco filled two needs; we don't have wine to sell, and I wanted very badly to make a less expensive wine. But even Rosso and Bianco can be $20 in a restaurant, which I hate."

Francis Coppola is sitting on the verandah of his family home, surveying his vineyards of Cabernet Sauvignon grapes, which will make grand wines redolent of what André Tchelistcheff called "Rutherford Dust." After fits and starts in the wine business, after almost losing this magnificent property, and after succeeding in a twenty-year dream to reunite the original Inglenook, Francis is talking about his future beyond Niebaum-Coppola.

"I think I'm going to turn my attention from this shortly. Once I feel it's on the right track I will look in other directions. Right now, I'm at a time in my life where there's a lot I want to do intellectually—read, study, with a view to making my own film. A personal film, a real film. I just now at this age have become almost reclusive, and I resent demands on my time.

"We have a big library, and I have some ambitious ideas related to making a film, and I want to study physics and economics and philosophy, and the stuff I want to think about."

It is evident that Francis Coppola, as he enters his sixties, is thinking about his place in the order of things, and on planes both theoretical and practical. It sounds like this gift of time to think is a gift that Coppola must give to himself. Many of us would like to breathe that rarefied air, but may never have the luxury of time to contemplate our place in the cosmos.

What is hard to understand about Francis freely turning his attention away from the lands of Niebaum-Coppola so that he may explore his own interior landscape is this: after fighting so hard and sacrificing so much to attain his dream, his vision, isn't the old Inglenook estate too precious to him to not make it the center of his life?

Speaking to the question with what sounds like the voice of a man who has, indeed, been taking the first steps in a long-range expedition to the world beyond the material, to the world of ideas, Francis Coppola takes a big puff of his small cigar, exhales and answers:

"I love this place as I have always loved this place. The laws of emotion are constant. Happiness is Happiness. It's like relativity or the speed of light. When he's happy, the poorest man in the world is just as happy as the richest man in the world. When the richest man in the world is happy, he isn't any more happy than the poorest man in the world."

Tasting Notes

The Rubicon Vintages from 1978–1995

Wines rated on a 100 point scale:

95–100	*extraordinary*
90–94	*excellent*
85–89	*very good*
80–84	*good*
75–79	*average*
70–74	*below average*
65–69	*poor*

Hold	*a wine to keep*
Drink	*enjoy now*

1995
(released in bottle on March 15, 1999)

Drink/Hold

Barrel Sample tasted October 7, 1997
Finished Wine tasted on its day of release, March 15, 1999

I did not think it possible that the 1995 Rubicon could be as good or better than the 1994, but winemaker Scott McLeod has proven me wrong. I am truly happy to be wrong, because this is The Perfect Rubicon, one of the best wines I have ever tasted in my life.

The color is just shy of inky black, with just the slightest touch of red/maroon highlights. The nose is redolent of blackberries, black currants, black cherries, ripe plums, fresh fig, just a touch of eucalyptus and oregano, and the ascendant creamy vanilla tones and aroma of new French oak. On the palate, the wine is surprisingly soft and supple, leading with a sweet and luscious attack of jammy blueberries and plums, with a touch of mint and black pepper in the background. Tannins are mouthfilling but mostly soft, balanced with the compelling fruit and refreshing acidity of Cabernet Sauvignon (88%), and the bright red fruit flavors of Merlot

(6%) and Cabernet Franc (6%). The wine has no rough edges, and is heat-seeking its way to achieving palatal harmony. Incredibly nuanced and elegant even at this young age, this is a wine to drink now and twenty years from now. A masterpiece, reveling in the sunshine of the Napa Valley and the balanced flavors and historic pedigree of the Niebaum Clone of Cabernet Sauvignon from the estate. Much like the 1994, the finish of this wine goes on virtually forever, changing on the palate even as the wine changes in the glass. Consistent notes on barrel sample and bottle sample, indicating a wine that is still in its infancy, a wine that will last and live a long and healthy life. An incredible feat by winegrower/winemaker McLeod, following on the heels of the near-perfect 1994 Rubicon. 4,565 cases produced; retail $80.

Author's rating: **100**

1994
(*released on March 15, 1998*)

Drink / Hold

A wine forever etched on my palate, a classic Rubicon from a classic vintage. The 1994 Cabernets from the Napa Valley are among the finest wines ever produced in California, and this benchmark Rubicon (about 92% Cabernet Sauvignon) with its luscious complexity will make the reputation of Rubicon and winemaker Scott McLeod. Tasting the wine is a new experience for the senses, as they are almost overwhelmed by its deep black color streaked with cherry red, its nose of plum, red currants, and blackberries enfolded in rich but balanced oak. The flavor is all about spice, length, density, harmony, and ripe, sweet fruits that speak volumes about the unique character of Rubicon's earthly components. A bit more subtle, a bit more restrained, a bit more elegant than the 1993, and pleasingly so. This wine redefines the signature of the Niebaum-Coppola vineyards and of Rubicon. A truly singular wine for now and for the ages, delivering a complex whirl of synergistic and sensual pleasure upon pleasure with each sip. A near-perfect wine, eclipsed only by the 1995 Rubicon.

Author's rating: **98+**

1993
(*released in 1997*)

Hold

A very good wine from a difficult vintage due to uneven ripening of red fruit in many Napa vineyards, including the Niebaum-Coppola estate. Color is quite dark, with extracted tannins giving the wine body and depth. The nose reflects black cherries and cassis, notes of raspberries, and a bit of pine cone tarriness, probably due to a more extended maceration. The flavors are big, rich, typical Cabernet Sauvignon—tannins balanced by black cherry fruits, a touch of eucalyptus, and surprisingly refreshing acidity. This wine lacks some of the forward fruit of the 1994 and 1995 vintages, and needs time to achieve the unison of flavors that it seeks. Released in 1997, it is best to drink from 2002 and beyond.

Author's rating: **90**

1992
(*released in 1996*)

Drink / Hold

From an almost-perfect growing season without rain, the wine shows how well the dry-farmed vines of the Niebaum Clone of Cabernet Sauvignon respond to heat and to stress. A tremendous concentration of black and red fruit flavors already in harmony, with a nose of oak-tinged raspberries, and a deep black cherry color that heralds longevity. I love the chewy density of the sweet and bittersweet tannins, making for a rich wine with real "grip"; a signature Rutherford Cabernet Sauvignon (about 93%) with backbone and structure that still delivers pleasure at this young age and for the next fifteen to twenty years.

Author's rating: **93**

1991
(released in 1996)

Drink / Hold

A watershed year. Scott McLeod's first Rubicon vintage brings about a radical stylistic change in the wine, which was hinted at in the 1990 vintage (see next), which McLeod finished for the departing winemaker, Steve Beresini. The wine is 90% Cabernet Sauvignon, full of inky black richness and chewy complex structure for the long haul, all balanced by a nose and flavor of sweet black cherries provided by the luscious Merlot and Cabernet Franc grown on the estate. The first Rubicon to really celebrate its base material, the wonderful fruit. Taste the black currants, anise, allspice, and just a restrained touch of cedar astringency. Compared to earlier wines, the '91 Rubicon has much more density and heft on the palate, and finally the tannins are not overly extracted, so that its youth is charming, not alarming, on the palate. This Rubicon has all its ducks in a row, and all singing harmony, flying gracefully over the Daniel/Inglenook vineyards that, with this vintage, receive the historic imprimatur of McLeod/Niebaum-Coppola, and begin to tell of wines yet to come.

Author's rating: **95**

1990
(*released in 1995*)

Drink/Hold

Nineteen ninety, 1994, 1995 and probably 1997 were the four best growing years of the '90s in the Napa Valley, and the 1990 Rubicon is a fine wine, reflective of the vintage at harvest. Ripe fruits, especially plums, blackberries, and black and red currants jump out of the glass, and a tannic background of black licorice/anise, cinnamon, black pepper, and tar in the finish promise a wine that will deliver youthful pleasures but age gracefully. A beautifully balanced wine. What I really like is the downplaying of oak as a major component in the nose and flavor of the wine. The last Rubicon made by Steve Beresini, the wine was finished in the barrel by Scott McLeod.

Author's rating: **94**

1989
(*released in 1994*)

Drink

A rare disastrous vintage in most of the central Napa Valley, with lots of rain during harvest, following on the heels of drought-driven 1988, another poor vintage. Crops were large and grapes ripened unevenly, with the rain diluting flavors overall, especially in Cabernet Sauvignon. The '89 Rubicon looks and smells promising, but is a very thin wine on the palate, with hard tannins that, in the tradition of Bordeaux wines from poor vintages, will probably never soften up.

Author's rating: 76

1988
(*released in 1993*)

Drink

El Niño brings a drought and heat wave to the valley, leading to small crop yields, berry shatter, and uneven ripening, especially for Cabernet Sauvignon, and this wine is 100% Cab. The nose of the '88 Rubicon is musty and reminiscent of raw unripe beets without a lot of fruit, red or black. On the palate the wine is so astringent that it is painful to taste eleven years after vintage. 1988 and 1989 (see above) are the Rubicon vintages to avoid, a truism throughout the universe of Napa Valley Cabernets.

Author's rating: 74

1987
(released in 1993/1994)

Drink / Hold

The drought of 1987, with a warm but not brutal growing season, stressed the vines just enough to produce a very small crop of very ripe grapes. The color of the wine is vibrant red, almost black cherry, and the nose is of sweet red beets, sour cherries, and leather. The wine tastes of red fruits, especially strawberries, raspberries and cherries, but it lacks real cool weather–driven acidity. It has a slight taste of acidulation, kind of like a baby aspirin in the finish, which will mostly disappear when enjoyed with food. Overall, the wine is balanced, and is a very good example of what Napa's Oakville/Rutherford Cabernet axis was producing in the late 1980s, a far cry from the jammy fruit-driven wines of the late '90s.

Author's rating: **87**

1986
(released in 1992)

Hold

This Rubicon represents the pinnacle of the Bordeaux style of Napa Cabernet promulgated by André Tchelistcheff, which is not to say that this is not a wonderful wine; it is a fine wine from an excellent vintage. The '86, however, is unrecognizable as a modern Rubicon, and is a hefty, complex wine, still black cherry in color, with a nose of plums, fresh stone fruits, and, at least for Rutherford Cabernet Sauvignon, uncharacteristically prominent eucalyptus and menthol. It has a certain earthiness that is attractive, if not elegant. The wine is quite astringently tannic, but balanced by oak, not fruit. This is a wine of great structure modeled on the Old World, a wine that is not yet ready to drink, but is getting close. Hold until at least 2003–2008.

Author's rating: **90**

1985
(*released in 1991*)

Drink

1985 is considered the vintage of the decade in the Napa Valley, especially for Cabernet Sauvignon; so what happened to Rubicon? This is the most blatant example of interventionist winemaking that I have found in this tasting. It's amazing that this wine didn't sink Rubicon's fortunes for all time, as it showed none of the finesse inherent in the vintage, except in the color of the wine, which is vibrant red. Who made the decision to put more than 25% Merlot in a wine that should have sung the praises of the 1985 Cab? The nose is full of barnyard, vegetal, sulfurous, off-putting aromas, and has an unpleasant smoky char. The flavor is diluted, not rich. This wine makes me want to cry when I think about the exquisite Cabernet fruit that must have been grown on the estate in this year. I keep thinking about the "smoker's palate" of André Tchelistcheff, who was the consultant on this wine, and who was never without a cigarette, even during wine tastings. A wine for smokers only, maybe for one of those ubiquitous cigar dinners. A just-average wine from a magnificent vintage, which makes it a big loser.*

Author's rating: 77

---

* I should point out that many other wine writers do not agree with this assessment. For example, in 1989, the well-respected *Wine Spectator* senior editor James Laube, on page 278 of his book, *California's Great Cabernets*, rated the wine, which had not yet been released, 91 points out a possible 100, writing that the wine was "An elegant, delicately proportioned 1985 that is just opening up, with crisp acidity, firm structure and

1984
(*released in 1990*)

Drink

Once again, an excellent vintage, but the Rubicon did not show all that much. At least, the wine was very good, unlike the '85, but it should have been excellent with the prime growing conditions of 1984. The color is black cherry, but beginning to change from opaque to translucent; the wine is ready to drink. In the nose, there are cherries, blackberries, and tar. While light in body, the wine has a certain earthiness that gives the slightest suggestion of complexity without much tannin. Again, like the '85, this Rubicon does not reflect its history and *terroir*; it is a winemaker's wine, a wine of intervention, but with better results. In the finish, the wine is lean and acidic without much depth, but there is something pleasant about the bittersweet flavors of chocolate and cherries. Although it is a bit too delicate for a classic Rubicon, it is a good food wine. It will not overwhelm any but the lightest dishes, and is content to be a backgrounder for rich foods.

Author's rating: **86**

---

lively cherry, plum, vanilla and currant flavors that offer intensity and depth. The finish is long and smooth. Should be a beauty. Drink 1995–2003." In 1995, on page 413 of his book, *California Wine,* he downgraded the wine somewhat, to 88 points, writing that the wine is "Tannic and concentrated, with earthy, spicy aromas and ripe plum and anise flavors that pick up a mineral edge." Both of Laube's very fine and comprehensive books are published by the Wine Spectator Press.

There was no 1983 Rubicon.

1982
(*released in 1988*)

Drink / Hold

A difficult vintage, with rain at harvest. A very pretty color, medium black cherry. Consistent with the '84 (see above), the nose is all cherries, blackberries, and tar, but with far more body, uncharacteristic of this watery vintage. Both the acid and the tannin are still ascendant in this wine, and we can only hope that a softer harmony is in its not-too-distant future. Just misses a Cabernet signature. A hard wine to judge, harder to predict, but it still needs time.

Author's rating: **86**

1981
(*released in 1987*)

Drink / Hold

A very earthy wine, typical of a good Cabernet vintage. The color is dark purple, and the wine is beginning to open, judging by color and nose. Quite earthy and alcoholic, with warm Mediterranean notes on the palate, the '81 Rubicon is a pleasant surprise, with its big tannins but even bigger juicy red fruits. Not quite ready, but this wine provides a fabulously earthy but elegant counterpoint to rich foods. Very much its own wine, it seems unrelated to any other Rubicon from any other vintage. A good example of 1981 Napa Cab, but is it a good example of Rubicon?

Author's rating: **88**

1980
(*released in 1986*)

Drink

Back to Bordeaux. Nineteen eighty was the first year that André Tchelistcheff consulted on the winemaking at Niebaum-Coppola. This wine is dried out and tannic, never having reached its full potential. Not atypical of this vintage year, which had several heat waves running through the Napa Valley, realizing grapes that were ripe but unbalanced. The finished wine seems to have had massive amounts of acid added, making it clumsy and awkward. There are quite a few people who enjoy this kind of wine; many of them live in Great Britain. About ten years past its prime, when it probably would have been a very fine Bordeaux-style California Cab.

Author's rating: **80**

1979
(*released in 1985*)

Drink

This wine is still opaque, and is just now starting to fade to brown, telling me that it never reached balance in the glass. A rustic wine, the nose smells of mushrooms, earth, manure, soil, which might just be brettanomyces, a wild yeast found in the cellar that can foul the nose, flavor, and longevity of a wine. On the palate the wine has a certain gaminess that is unpleasant. Not a successful wine, especially in light of the first Rubicon, made in 1978 (see below).

Author's rating: 72

1978
(*released in 1984*)

Drink

A perfect year to make the first vintage of Rubicon, as the 1978 vintage is second only to 1974 for well-defined Cabernet, and the demand for the wine was great, following the Paris Tasting of 1976 by only two years. This is a great Rubicon, one of the best ever produced, and while it may now just begin to show some browning on its rim, it is in perfect condition for drinking now and for the next five years. The nose is redolent of raspberry essence, strawberries, tar, and a bit of charcoal. The wine is complex, supple and rich, the fruit is still fresh, and the tannins are soft and ripe. The winemaker for the 1978 and 1979 vintages was Russ Turner. The 1978 Rubicon is a classic, with its sweet finish, which shows the breed of the vines and their luscious fruit expressed on this historic property. A very promising start for America's second Meritage* wine.

Author's rating: **93**

---

* The first Meritage was Insignia, made by Joseph Phelps starting in 1975. Opus One followed Rubicon in 1979. Meritage is a blend of the classic grapes of Bordeaux, but in an American wine. For red wines, Meritage is any combination of Cabernet Sauvignon, Merlot, Cabernet Franc, Petit Verdot, and/or Malbec. For whites, it's a blend of Sauvignon Blanc and Sémillon.

Bibliography

Adams, Leon. *The Wines of America.* New York: McGraw-Hill, 1990.

Allegra, Antonia. *Napa: The Ultimate Winery Guide.* San Francisco: Chronicle Books, 1993.

Amerine, Maynard. *Table Wines: The Technology of their Production in California.* Berkeley: University of California Press, 1951.

———. *The Technology of Wine Making.* Westport, Conn.: Avi Publishing, 1980.

Amerine, Maynard, and M. A. Joslyn. *Table Wines: The Technology of Their Production.* 2nd ed. Berkeley: University of California Press, 1970.

Amerine, Maynard, and Edward B. Roessler. *Wines: Their Sensory Evaluation.* 2nd ed. San Francisco: W. H. Freeman, 1983.

Anderson, Stanley, and Raymond Hull. *The Art of Making Wine.* New York: NAL-Dutton, 1991.

Asher, Gerald. *On Wine.* New York: Random House, 1982.

———. *Vineyard Tales.* San Francisco: Chronicle Books, 1997.

Bespaloff, Alexis. *The New Frank Schoonmaker's Encyclopedia of Wine.* New York: Morrow, 1988.

Blue, Anthony Dias. *American Wine*. New York: Doubleday, 1985.

Broadbent, Michael. *The Simon and Schuster Pocket Guide to Wine Tasting*. New York: Simon and Schuster, 1988.

———. *The New Great Vintage Wine Book*. New York: Knopf, 1991.

———. *Michael Broadbent's Guide to Wine Vintages*. New York: Simon and Schuster, 1993.

Clarke, Oz. *Oz Clarke's Encyclopedia of Wine*. New York: Simon and Schuster, 1994.

———. *Microsoft Wine Guide*. Seattle: Microsoft Corporation, 1995. CD-ROM.

———. *Oz Clarke's Wine Advisor*. New York: Simon and Schuster, 1998.

Conaway, James. *Napa*. New York: Avon, 1992.

Daniel, John, Jr. "Notes on the History of Napa County Viticulture and Winemaking." Speech delivered on November 23, 1969. Recorded in *History of Napa Valley: Interviews and Reminiscences of Long-Time Residents*, vol. 1. St. Helena: Napa Valley Wine Association, 1974.

Darlington, David. *Angels' Visits: An Inquiry into the Mystery of Zinfandel*. New York: H. Holt, 1992.

De Groot, Roy. *The Wines of California, the Pacific Northwest, and New York*. New York: Summit Books, 1982.

Ensrud, Barbara. *American Vineyards*. New York: Random House, 1990.

Fisher, Mary Frances Kennedy. *The Story of Wine in California*. Berkeley: University of California Press, 1962.

Ford, Gene. *The French Paradox: Drinking for Your Health*. San Francisco: Wine Appreciation Guild, 1993.

Galet, Pierre. *A Practical Ampelography*. Translated by Lucie Morton. Ithaca: Cornell University Press, 1978.

George, Rosemary. *The Simon & Schuster Pocket Wine Label Decoder*. New York: Simon and Schuster, 1989.

———. *Lateral Wine-Tasting*. London: Trafalgar, 1993.

Halliday, James. *The Wine Atlas of California.* London: Viking Penguin, 1993.

Haraszthy, Arpad. *Wine-Making in California, 1871–72.* San Francisco: Book Club of California, 1978.

Impact Databank Report: The International Wine Market. New York: M. Shanken, 1998.

Johnson, Frank. *The Professional Wine Reference.* New York: Harper & Row, 1983.

Johnson, Hugh. *Wine.* New York: Simon and Schuster, 1987.

———. *Vintage: The Story of Wine.* New York: Simon and Schuster, 1989.

———. *How to Enjoy Wine.* New York: Simon and Schuster, 1990.

———. *The World Atlas of Wine.* 4th ed. New York: Simon and Schuster, 1997.

———. *Hugh Johnson's Modern Encyclopedia of Wine.* New York: Simon and Schuster, 1998.

———. *Hugh Johnson's Pocket Encyclopedia of Wine.* New York: Simon and Schuster, 1998.

Johnson, Hugh, and James Halliday. *The Vintner's Art: How Great Wines are Made.* New York: Simon and Schuster, 1992.

Johnson, Robert. *The Consumer's Guide to Organic Wine.* Lanham, Md.: Rowman & Littlefield, 1993.

Kolpan, Steven, Brian H. Smith, and Michael A. Weiss. *Exploring Wine: The Culinary Institute of America's Guide to the Wines of the World.* New York: Wiley, 1996.

Lapsley, James T. *Bottled Poetry.* Berkeley: University of California Press, 1996.

Laube, James. *California's Great Cabernets: The Wine Spectator's Ultimate Guide for Consumers, Collectors, and Investors.* San Francisco: Wine Spectator Press, 1990.

———. *Wine Spectator's California Wine.* New York: Wine Spectator Press, 1995.

Lembeck, Harriet. *Grossman's Guide to Wines, Beers, and Spirits*. 7th ed. New York: Scribner's, 1983.

Lichine, Alexis. *Alexis Lichine's New Encyclopedia of Wine and Spirits*. New York: Knopf, 1987.

Lipp, Martin R., and David N. Whitten. *To Your Health: Two Physicians Explore the Health Benefits of Wine*. San Francisco: Harper, 1994.

Maresca, Tom. *Mastering Wine: A Learner's Manual*. New York: Grove-Atlantic, 1992.

———. *The Right Wine*. New York: Grove-Atlantic, 1992.

Matthews, Thomas. *A Village in the Vineyards*. New York: Farrar, Straus & Giroux, 1993.

McCoy, Elin, and John F. Walker. *Thinking About Wine*. New York: Simon and Schuster, 1989.

McGee, Harold. *On Food and Cooking*. New York: Scribner's, 1984.

Meyer, Justin. *Plain Talk About Fine Wine*. Santa Barbara: Capra Press, 1989.

Muscatine, Doris, Maynard A. Amerine, and Bob Thompson. *University of California/Sotheby Book of California Wine*. Berkeley: University of California Press and Sotheby Publications, 1984.

Novitski, Joseph. *A Vineyard Year*. San Francisco: Chronicle Books, 1983.

Olson, Steve. *Food & Wine's Wine Tasting*. New York: Times Mirror Multimedia, 1994. CD-ROM.

Parker, Robert. *Wine Buyers Guide*. New York: Simon and Schuster, 1998.

Perdue, Lewis. *The French Paradox & Beyond: Live Longer with Wine and the Mediterranean Lifestyle*. California: Renais, 1992.

Peterson, Richard G. Interviews with André Tchelistcheff (1972, 1975), in *History of Napa Valley: Interviews and Reminiscences of Long-Time Residents*, vol. 2. St. Helena: Napa Valley Wine Association, 1979.

Peynaud, Emile. *Knowing and Making Wine*. New York: Wiley, 1984.

———. *The Taste of Wine*. San Francisco: Wine Appreciation Guild, 1987.

Pinney, Thomas. *A History of Wine in America: From the Beginnings to Prohibition.* Berkeley: University of California Press, 1989.

Prial, Frank, ed. *The Companion to Wine.* New York: Prentice Hall, 1992.

Ray, Cyril. *Robert Mondavi of the Napa Valley.* London: Heinemann/Peter Davies, 1984.

Robinson, Jancis. *The Great Wine Book.* New York: Morrow, 1982.

———. *Vines, Grapes, and Wines.* London: Mitchell Beazley, 1986.

———. *Vintage Timecharts: The Pedigree and Performance of Fine Wine to the Year 2000.* New York: Weidenfeld & Nicolson, 1989.

———. *Oxford Companion to Wine.* Oxford: Oxford University Press, 1994.

Roby, Norman, and Charles Olken. *The New Connoisseur's Handbook of California Wines.* New York: Knopf, 1993.

Simon, André. *Wines of the World.* New York: McGraw-Hill, 1972.

Simon, Joanna. *Discovering Wine.* New York: Simon and Schuster, 1995.

Skoda, Bernard. Interview with George Deuer, in *History of Napa Valley: Interviews and Reminiscences of Long-Time Residents,* vol. 2. St. Helena: Napa Valley Wine Association, 1979.

Steinberg, Edward. *The Vines of San Lorenzo.* Hopewell, N.J.: Ecco Press, 1992.

Sterling, Joy. *A Cultivated Life: A Year in a California Vineyard.* New York: Random House, 1993.

Stevenson, Robert Louis. *The Silverado Squatters.* London: Stone and Kimball, 1895.

Stevenson, Robert Louis. *Napa Wine.* San Francisco: Westwind Books, 1974.

Stevenson, Tom. *The New Sotheby's Wine Encyclopedia.* New York: DK Publishing, 1998.

Sullivan, Charles L. *Napa Wine: A History.* San Francisco: The Wine Appreciation Guild, 1992.

Sutcliffe, Serena, ed. *Great Vineyards and Winemakers.* New York: Routledge, 1981.

Taylor, Ronald B. *Chavez and the Farm Workers*. Boston: Beacon Press, 1975.

Teiser, Ruth. *California Wine: A Series of Oral Histories*. Interviews by author with: Leon Adams (1972, 1990), Maynard A. Amerine (1969, 1971), Legh F. Knowles (1990), Robert Mondavi (1984), André Tchelistcheff (1979, 1983), Albert Winkler (1970, 1972). San Francisco: Bancroft Library, 1969–1990.

Teiser, Ruth, and Catherine Harroun. *Winemaking in California*. New York: McGraw-Hill, 1983.

Thompson, Bob. *Notes on a California Cellarbook*. New York: William Morrow, 1988.

———. *Simon & Schuster Pocket Guide to California Wines*. New York: Simon and Schuster, 1990.

———. *The Wine Atlas of California, with Oregon and Washington*. New York: Simon and Schuster, 1993.

Unwin, Timothy. *Wine & the Vine*. London: Routledge, 1991.

Wait, Frona Eunice. *Wines and Vines of California*. San Francisco: Bancroft Co., 1889.

Wagner, Philip. *Grapes Into Wine: The Art of Winemaking in America*. New York: Knopf, 1976.

Waugh, Alec. *Wines and Spirits*. New York: Time-Life, 1968.

Wine Spectator Press Staff. *The Wine Spectator's Annual Wine Buying Guide*. New York: M. Shanken, 1998.

Winkler, A.J. *General Viticulture*. Rev. ed. Berkeley: University of California Press, 1974.

Zraly, Kevin. *Windows on the World Complete Wine Course*. New York: Sterling, 1999.

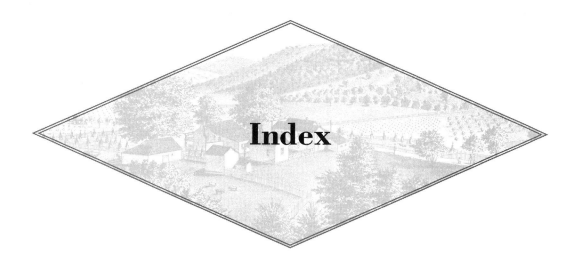

Index